D1546747

CONTEMPORARY RANCHES OF TEXAS

NUMBER NINETEEN

The M. K. Brown Range Life Series

Contemporary Ranches of Texas

BY LAWRENCE CLAYTON

PHOTOGRAPHS BY WYMAN MEINZER

University of Texas Press, Austin

Printed in Hong Kong
First edition, 2001

♾ The paper used in this book meets the minimum requirements of ANSI/NISO Z39.48-1992 (R1997) (Permanence of Paper).

LIBRARY OF CONGRESS CATALOGING-IN-PUBLICATION DATA

Clayton, Lawrence, 1938–
Contemporary ranches of Texas / by Lawrence Clayton ;
photographs by Wyman Meinzer.—1st ed.
p. cm. — (The M.K. Brown range life series ; no. 19)
Includes bibliographical references.
ISBN 0-292-71239-1 (cloth : alk. paper)
1. Ranches—Texas. 2. Ranching—Texas. 3. Ranch life—Texas.
I. Title. II. Series.

SF196.U5 C63 2001
636´.01´09764—dc21 2001027789

PAGE I: *Original saddlehouse, Long X Ranch*
PAGE III: *Gage and Tom Moorhouse*

Contents

Acknowledgments

I WANT TO EXPRESS MY APPRECIATION TO THE FOLLOWING individuals who have provided helpful suggestions, interviews, or responses to the material or who made suggestions of which ranches to include: Jerald Underwood of Uvalde; Bruce Cheeseman of Kingsville; Curt and Katy Hoskins of Van Horn; Mary Fru Reynolds Pealer and Mary Joe Reynolds of Kent and Fort Worth; members of the Jones family (especially W. W. Jones III, Dick Jones, and Kathleen Jones) of Hebbronville and Corpus Christi; Frank and Joy Graham of Alta Vista Ranch near Hebbronville; members of the Brown family (especially Rob and Peggy) of Throckmorton; Dolph Briscoe, Jr., and Charles Brown of Uvalde and Carrizo Springs; Clayton Williams, Jr., of Midland; Ted Gray of Alpine; Claudia Ball of San Antonio and Hudspeth River Ranch; George and Sue Peacock of Albany; Louise O'Connor, Kai Buckert, Joe Keefe, and Leah Bianchi of Victoria; Bob Reagan of Three Rivers; Marion Bassham and Bob Northcutt of Colorado City; Robert Waller of Albany and Mary Frances "Chan" Driscoll of Midland; Tom Moorhouse of Benjamin; Gus T. Canales of Premont; and Darrell Cameron, Lee Palmer, and Jay O'Brien of Amarillo. Without the support of these individuals and their cowboys and vaqueros, this book would not have come into existence.

Appreciation for the continuing support of my wife, Sonja, and to my two daughters, De Lys Mitchell and Lea Clayton, is recognized here as well.

The computer assistance given by Ted Paup is also gratefully acknowledged, as is the able support of my administrative assistants, Ann Giddens and Peggi Gooch, and Hardin-Simmons University.

CONTEMPORARY RANCHES OF TEXAS

Introduction

IN THE LATE SIXTEENTH AND EARLY SEVENTEENTH centuries, the Spanish Conquistadors introduced the system of ranching to the New World when they brought horses, cattle, sheep, and goats as part of their expeditions—horses to ride; sheep, goats, and cattle for food, leather, and tallow. They also brought people to tend the stock. Because in colonial times the Spanish did not castrate most male animals, every animal they brought was, in effect, breeding stock, which spread throughout the areas into which Spanish colonization efforts extended. Among their various enterprises, ranching was a major focus in order to feed the workers. In areas where mining developed, hides provided bags for ore, and tallow rendered from animal fat furnished candles to light the mines. No outside market existed for beef. The meat was the least important part of the animal.

The earliest ranches were situated on huge land grants that were at first awarded to supporters of the Spanish king and later, after Mexico gained independence from Spain in 1821, to those loyal to the Mexican government. Some of these huge estates extended far into Texas. On these "open-range" ranches were established the practices by which cattle were worked.

An open-range ranch usually consisted of a herd of cattle, a crew of mounted herders, and a stretch of grazing land. The ranch was as mobile as necessary to keep the cattle supplied with good grass. Cattle were gathered in the spring for working the calf crop, but were otherwise left to roam the grasslands.

Perhaps the most important contribution of the Spanish system was the introduction of mounted herders—the vaqueros—the men who worked with cattle. The vaquero used a rope (the *riata* or *lazo*) and developed protective leather leg coverings today called chaps. He contributed as well to the development of saddles suitable for working cattle.

In the 1830s and 1840s with the influx of immigrants into Texas from the southern United States came people familiar with cattle raising in Europe and Britain. Some of these people had been raising livestock in their home states, and many pursued cattle ranching in Texas, where they picked up new techniques for ranching on the plains—techniques learned largely from the Mexican vaquero.

Just after the Civil War came the expansion of trail driving, perhaps the most important development of this time. The mounted herder tending cattle on pasture assumed a new role—that of drover. The practice brought immense herds of cattle to railheads in Kansas, where they were connected with markets in the eastern United States hungry for beef. In addition, the vast western grasslands had been stripped bare of native cattle (bison or buffalo) by the demand for hides, and there arose a need for Texas cattle to repopulate the range. The visibility offered by driving herds of 2,500 or more mature steers or mixed herds of Longhorn breeding cattle to railheads and ranges on the northern plains brought the cowboy to the attention of a world that soon mythologized him as the ultimate western frontiersman. The outfit or appearance associated with the cowboy developed during this period—tall hat, high-heeled boots, spurs, and the like.

The heyday of open-range ranching lasted until the mid-1880s. Several factors ended that kind of ranching. One of the most significant was climatic. From 1886 to 1888, cataclysmic weather, characterized by drought-ridden summers and incredibly cold winters, decimated cattle herds all across the West. The market was flooded as ranchers tried to cut their losses by selling stock before the animals starved or froze to death. Ranching went into a sharp decline.

When the enterprise emerged from this recession, ranchers knew that to survive they must breed more desirable kinds of cattle. In order to accomplish this goal, they fenced their properties to keep scrub bulls away from the cows. Barbed wire spread across ranges of the West. Ranchers also realized they had to provide feed for their stock during periods of cold weather and drought in order to see the herds through these critical times. Ranching was conducted very differently after what some called the "great die-up."

Also important to the development of the system of enclosed pastures were drilled wells and the use of windmills to pump water to the surface for watering stock. Hence, access to natural sources such as streams was no longer essential. In areas with no subsurface aquifer, the solution was to dam natural draws and scoop out earth to deepen these depressions, called in most of ranch country by the name *tanks*, not *ponds*. Teams of mules pulled fresnoes, scooplike devices, to move dirt. In areas without subsurface aquifers, this became the only water available for man and beast.

The bachelor cowboy was still prevalent during this period, living part of the time on the range around the chuck wagon and part of the time in the bunkhouse at the ranch. The line shack—usually a small house and corral with minimal supplies to shelter a cowboy and his mount away from the headquarters—continued to be important, only now cowboys were not pushing cattle back onto rangeland claimed by the ranch, but were checking and repairing fences as necessary. At first, Hereford and Durham cattle, English breeds with desirable characteristics for beef production, proved popular for crossbreeding Longhorn stock. Later, purebred herds evolved as well.

During this period, improvements in horse breeding also grew out of the availability of high-quality stallions in private ownership and at government breeding stations. The principal aim was to provide good remounts for cavalry forces. The effort, nevertheless, improved ranch horses.

As the railroad system expanded into cattle-producing areas, with it came farmers and other settlers, a development that wrought incredible change to life in ranch country. Despite the expansion of the railroads, however, trail drives continued until almost the end of the 1800s because it was cheaper to move cattle that way than to pay the expense of hauling by train. Large crews of cowboys were still required for this work. In the first half of the twentieth century, after the era of trail drives ended, ranching gradually modernized as transportation improved. A common technique cattlemen used was to haul cattle to the railroad shipping point by truck. For many ranchers, this eliminated one- or two-day drives that could reduce the weight of their stock.

The next major change occurred as a result of World War II. Most of the young men were drawn into the war effort, so in order to survive, ranchers relied on wives, daughters, old men, and those unfit for military service. The role of machinery escalated during this time with the advent and widespread use of such tools as working tables to hold the calves to be castrated, vaccinated, or treated for disease. A major advance was the availability of stock trailers, particularly those for hauling horses on the ranch. This allowed ranchers to get by with fewer men and fewer horses. One reason for the proliferation of trailers was the spread of the skill of welding, which some men and women had learned in the ship yards during the war effort and others learned afterwards as a result of G.I. Bill benefits at local colleges and trade schools. The trailer behind the four-wheel-drive pickup became a mainstay of the ranching industry. During this time many ranchers parked their chuck and feed wagons and started using pickups to haul their cowboys and feed around as well as to haul food to the cowboys, particularly for the noon meal. All these developments resulted in a need for fewer and fewer cowboys.

Cattle breeders improved their herds by importing the Brahman breed from India, as well as by continuing to depend on two British breeds, Hereford and Angus. The Brahman tolerated the heat, humidity, and parasites common in South Texas, although in Northwest Texas and the Trans-Pecos, this immunity was less of an advantage. King Ranch developed the Santa Gertrudis breed, the first breed of cattle developed in the United States and recognized as such in 1940. The ranch later developed the Santa Cruz breed as well.

Horse breeding also continued to play an important role in Texas ranching culture. After World War II, the establishment of the American Quarter Horse Association stimulated the breeding of these horses characterized by strong hindquarters and a blocky build combined with quickness and cow sense. Many stockmen considered Quarter Horses to be the best ranch horses, and they established excellent breeding programs built upon famous sires and high-quality mares. At midcentury most ranches still carried six to ten horses per man rather than the eight to twelve that had been common during earlier years. Cowboys often lament that from this time on, however, because horses were hauled and not ridden those long miles to and from work, the animals' level of experience has never matched its pre-war excellence.

Other changes came to ranching as well. For instance, with the eradication of the screwworm, managing a herd of cattle involved less and less daily prowling of pastures to rope and doctor infested animals. In addition, the presence of local auction rings allowed small ranchers to haul their own cattle to market, though the important regional markets in such places as Fort Worth and San Antonio continued to attract large ranchers before finally declining. The Fort Worth market has become a thriving historical district that draws on the mystique of earlier days. Furthermore, many ranches closed down their bunkhouses and relied instead upon a series of "camps" with houses, barns, and pens. The camps were strategically placed around the ranch and manned by married men, whose families lived with them in these sometime remote areas. Ranch families experienced more and more contact with town life as progress brought electricity, telephones, running water, and improved transportation into their lives.

The post-war era also saw attempts to eradicate the brush that had long been encroaching in South Texas and had become a problem on ranches throughout the state. By using long, very heavy chains stretched between two bulldozers, ranchers began knocking the brush down to improve the quality of grazing. The bulldozer continues to play an important part in water conservation in constructing surface tanks to catch and hold runoff.

The use of cross-fencing (to make pastures smaller and more manageable) expanded, but the most significant development in grazing practices stemmed from the work of Allen Savory, a pioneer in the field. The Savory Cell Grazing method involves fencing pastures into small plots, often a hundred acres or less, putting more cattle on a plot than the land can support, and keeping the stock on the plot for only a short period (a few days or weeks). The animals eat even the undesirable growth, not just the best, and they crush dried weeds and coarse grasses to powder. They also fertilize the plot with manure. After a few days or weeks, the cattle are removed for an extended period to allow the land to recover fully and become even more productive.

Today, technology plays an important role in ranching. Many ranchers use computers to track the performance of their animals. Medical practices have improved to include more vaccinations and medications to treat sick animals and prevent diseases. Brucellosis, a bacterial disease that was long a threat to breeding herds, has been virtually eliminated. Ranchers have learned to palpate cows that have been bred to see if they have actually conceived. Therefore, the ranch does not waste money feeding a cow that is not going to produce a calf that year.

Other technology has played a prominent role in ranching as well. One technique gaining ground is the use of imaging devices to determine the quality of the animal's carcass while it is still alive. Another is the use of videotape to market cattle. Stock sold this way does not have to be gathered and sent to a market but can be shown to potential buyers over a video network, purchased, and then gathered and shipped in large cattle vans straight to a feedlot or pasture, thus avoiding stress on the animals and preventing weight loss. Some ranches have adopted a ranch-to-rail ownership pattern in which the ranch retains ownership through the feeding out and slaughtering of the animal. In that way, the rancher has a chance to make a double profit, still often not a major margin.

Ranching continues to change. Individual ranches have tended to become smaller. Estates are divided following the deaths of previous generations. Sometimes land has to be sold outright to pay heavy taxes imposed on inherited estates. Crossbreeding of cattle with European, African, and Asian breeds has become popular, even to the point that the cattle have become too large to thrive on sparse ranges. Horses of excellent bloodlines continue to flourish. Many ranchers, however, choose to buy mature horses of four to six years rather than raise their own. Thus, they eliminate that sometimes expensive and time-consuming activity. This practice also reduces the need for cowboys to ride young, unruly horses, always a dangerous job but one of the traditional tasks of range life. Nig London, a Throckmorton County cowman, noted to me in an interview, "We used to have horses that would try to hurt you. It finally occurred to us it cost as much to feed one of those as it did a good horse, so we just got rid of those bad ones." The use of Quarter Horses for cutting contests and barrel racing—that is, for sport—has grown significantly.

Because ranchers hire fewer and fewer hands, the number of horses needed per man has shrunk to sometimes as few as two to as many as six. A common practice is "swapping help" with a neighboring ranch; that is, pooling help on given days, or hiring temporary cowboys, called "day workers." With the use of trailers, less riding time is required, and the two-to-six horses per rider is adequate for many ranches.

Modern life with air-conditioning, swimming pools, and the like has come to ranchers just as it has to people in town. Most cowboys these days are married. Some live in town and drive to the ranch each day, or they live in a camp on the ranch, which usually provides much better housing than it did thirty or forty years ago. Improvements to grazing land include periodic controlled burns to eradicate undesirable and parasitic growth. The use of chemicals to control brush, particularly mesquite and prickly pear, also has gained a great deal of favorable attention.

As ranch management becomes increasingly complex, ranch managers need more education. For large operations such as Spade Ranches, the general manager may have a doctorate, almost always from Texas A&M University, in one of the agricultural fields. Some managers are graduates of other excellent ranch management schools such those at Texas Christian University or Tyler Junior College.

From the time of the great Spanish estancias to the advent of the modern ranch, the methods and techniques of ranching have evolved, but indomitable spirit and deep resolve are still as vital to success as they were three centuries ago. As early ranchers struggled against nature and against other, often hostile, peoples, modern ranchers grapple with holding the ranch together despite various personal, economic, and climatic challenges.

THE SCOPE OF THIS WORK

This book tells the stories of sixteen working ranches in Texas, a state deservedly renowned for its contributions to the development of ranching. The story of each ranch begins with historical details of its founding. The sources of the material on the ranches discussed here are as varied as the history of each. Although the history of many ranches has never been formally recorded, some of this information can be found in printed form. Our generation owes a huge debt of gratitude to those who preserved those early stories and histories. The rest of the material in this book came from oral sources, always a fragile storehouse for facts, and often on the verge of disappearing. At best, the details of the founding of a ranch a century or more old are available only if the information has been passed down by oral tradition within the family. Not all of the ranches included here, however, have been in one family that long. Some have been sold over the years or have been split up in estate settlements over the decades and generations since founding, but they are today in the hands of people still intent on making them productive in the livestock business. Much of the land being ranched is suited for little else, because the better land—more fertile, better watered, and less sloped—has been converted to farming, and land near cities has fallen victim to urban sprawl.

In addition to relating their histories, I also have sought to include details of the current operation of these ranches, which represent different regions of the state. This information is available only by interviews and observation, and in a generation or two will constitute part of the historical record. Thus, it is "contemporary" history that will be valuable in the future.

I have tried as well to show the geographical diversity of this business in Texas. Ranching practices are strongly influenced by each region's unique features—amount of annual rainfall, depth and type of soil, types of flora, temperature range, and the like. A rancher's chances of survival depend on the ability to adjust to these factors and use them to advantage.

Brief discussions at the beginning of each section of the book describe in turn the three large regions included here. These regions—South, Panhandle and Northwest, and Trans-Pecos—are broadly drawn to accommodate the ranches discussed. The first section includes ranches from South Texas, here defined as that part of the state below a line stretching from Victoria on the east to Three Rivers in the central part of the state to Carrizo Springs in the west. The second region, Panhandle and Northwest Texas, includes the Rolling Plains and the lower part of the Panhandle, and lies north of Interstate Highway 20. The third, the Trans-Pecos, includes the Davis Mountains and the desert portions of West Texas along one of the most significant rivers in the state, the Pecos. All lie west of the ninety-eighth meridian, that imaginary line that marks the end of the timber belt stretching across East Texas and on eastward and the beginning of the plains. It is a line running near Fort Worth, Austin, and Corpus Christi. West of this line, ranching was one of the viable options for settlement and development because of the absence of timber, scant rainfall, and soils not always suitable for growing crops.

Some ranch operations do not fit neatly into these geographical boxes because several of them have property in more than one region. The important ranching culture in the Edwards Plateau, for example, is also not well represented here, because the Y O Ranch, which lies in this region, is included in my earlier volume, *Historic Ranches of Texas.*

Details of ranch life are important to this picture. Information on the type of cattle or other animals that graze these ranches is significant, as are the kind and number of horses used by ranchers to take care of the stock. The horse is still essential to ranching. I did not encounter a single ranch that uses motorcycles or all-terrain vehicles to herd cattle. However, late one afternoon I did witness a cowboy at the O'Connor Ranch use a four-wheel all-terrain vehicle to round up the horses so that the cowboys could have fresh mounts to use in the early evening to practice roping steers in an arena.

The details of the work vary: How the men work the animals, the life routine of the cowboys, and the gear that they use all receive attention. I include information on forage, parasitic brush, livestock diseases, and other issues unique to each region, because these factors determine the routine of life for those involved. I also discuss how the ranchers help the cattle through the winter months.

My primary focus is on cattle raising, though sheep and goats also range in some regions of the state, especially the Edwards Plateau, which encompasses the southern part of Central Texas between the Pecos and Colorado Rivers. The breeds of cattle preferred also vary regionally, for reasons explained in the story of each ranch. Some ranches raise horses, cattle, sheep, and goats; the determining factors are the kind of range available to the stock and the desires and preferences of the owners. If grasslands are prevalent, then cattle are likely preferred. If the range produces mainly weeds and brush, sheep and goats may be a wiser choice, but only if predators such as coyotes are not a threat. The test of a successful rancher is knowing what to do and what not to do, and when to act on a particular range. Ranchers who operate in several areas must know the strengths and weaknesses of each one. The ability to do what is right at the right time determines success; the inability to do so results in failure. An old adage warns that it is best to ranch in an area with which the rancher has been intimately familiar over a long period of time.

The South Texas ranches included in this volume are the Alta Vista Ranch near Hebbronville,

Canales Ranch near Premont, Catarina Ranch near Carrizo Springs, the Ray Ranch west of Three Rivers, and the O'Connor Ranch near Victoria. In the Panhandle and Northwest Texas are the R. A. Brown Ranch near Throckmorton, Chimney Creek near Abilene, the Goodnight near Claude, the JA near Canyon, the Moorhouse near Benjamin, the Nail Ranch north of Albany, and the Renderbrook Spade Ranch south of Colorado City. In the Trans-Pecos are Henderson Cove Ranch near Alpine, Hudspeth River Ranch north of Del Rio, Long X Ranch at Kent, and the 101 Ranch south of Van Horn.

Several criteria influenced the choice of these ranches instead of many other deserving operations. Those whose beginnings trace back to open-range days or at least back to the previous century were especially sought because of their historical interest. Prominence in ranching circles was a second influence on choice. People who ranch on a large scale frequently know each other and, more importantly, know the quality of other operations in the state. Effective management and vision attract attention and gain for ranches the reputation of running a quality "outfit."

Recognizability proved an important—though not overriding—factor because many significant and large ranches are not well known outside their own areas. Such names as King, Kleberg, Waggoner, Burnett, Swenson, Schreiner, and Goodnight are among the recognized names, and all but one of these I included in *Historic Ranches of Texas*. Many large, well-managed operations quietly and efficiently continue to operate in the state, all but unknown to nonranching families outside the immediate area of operation.

Historically interesting pasts ranked high in the criteria for inclusion because many of these ranches have fascinating stories connected with their founding and operation. For example, the Kuykendall family came to Texas with Stephen F. Austin's original Three Hundred, and descendants continue to operate the 101 Ranch brand established by one of their ancestors.

Fascinating personalities played a part as well. George and W. D. Reynolds, two of the most intriguing personalities in the settlement of the West, certainly merit attention in any discussion of ranching in Texas and beyond. So does Thomas O'Connor, whose vision of owning and fencing land led to his becoming one of the dominant forces in ranching in Victoria County. Claude Hudspeth, a legislator and rancher, also left a rich legacy to his descendants. Dolph Briscoe, Jr., of Uvalde was one of the dominant political figures of his day, and Clayton Williams, Jr., of Midland attracted a lot of attention in his 1990 bid to defeat Ann Richards and become governor of Texas. Both men continue strong ranching traditions. The J A and Goodnight ranches are legendary and could not be overlooked. The Jones family in South Texas has long avoided publicity but consented to be part of this project. The Renderbrook Spade Ranch has a strong claim for inclusion because of the family's background and the manner in which the owners continue to operate the ranch. These are but a few of the ranching people in Texas.

Geographic distribution was another factor. I wanted to show ranches from diverse areas of the state to illustrate operational variations while still retaining the similarity of background and purpose. In areas where the soil lends itself to farming and where rainfall is plentiful, such as in East Texas, the number of large ranches is small; settlement patterns cut the land into small tracts that would support a family at a time when small farms were a pattern of family life. Once broken up, these smaller acreages rarely were reassembled into large ranches. The largely treeless prairie beckoned to those with ranching dreams.

One consideration not to be overlooked is the willingness of ranch management to be included in the project. Ranch people, while often kind and generous, are private and shy of publicity, especially in these days when environmentalists campaign against their livelihoods. In the Davis Mountains, for example, inhabitants of the region perceive a strong federal effort to displace ranchers in order to return land to its natural state or to convert huge portions of it into parks. This has caused unrest among local ranchers and has encouraged them to shun attention.

Of interest to readers seeking a comparative study are the regional variations in ranching practices. From the lush coastal plain near Victoria where frost only occasionally kills the forage to the harsher, drier ranges of West and North Texas to the drought-prone ranges of South Texas, ranchers must respond differently to the needs of their stock. In southeast Texas and East Texas, such diseases as foot rot and river fever, as well as serious parasites, must be dealt with far differently from those in the drier climates of West Texas and the Davis Mountains. The kind of feed necessary for the winter, the most favorable time of year to work the calves, the number of acres needed to support an adult cow are among the many concerns that someone must make knowledgeable decisions about in order for the ranch to survive as a business entity. On the Coastal Plain near Victoria, a cow can thrive on three to five acres. In Northwest Texas, it needs twenty to thirty acres. In the West, the Davis Mountains, and on the 101 Ranch, each cow requires sixty acres or more. In the far western regions of the Chihuahuan Desert, these ratios often are stated in number of cattle per square mile or section of land rather than per acre.

How much land constitutes a ranch also varies by region. Ten thousand acres is a rule of thumb in Northwest Texas, because with this much acreage a rancher can have a horse pasture, a shipping pasture, and several pastures through which to rotate stock. Some ranchers devote at least part of their operation to pasturing yearling calves, which are bought, grazed for a time, and then sold. To be economically viable as a self-contained unit, a cow-calf operation must have at least three hundred cows. The average cow herd in Texas numbers fewer than fifty head. On most of the ranches included here, numbers run from several hundred to several thousand, depending on the range available to the rancher. Many of the small cow herds belong to people who support their herds and themselves with a job in town.

The financial commitment to ranching holds true regardless of the region. Buying protein supplements, minerals, hay, and other necessities cannot be put off until a convenient time. Payday for a ranch often comes only once a year when the calves go to market. Livestock must eat every day. Another range adage is that no one can starve a profit from a herd. In fact, each day a cow will eat about 2 percent of her weight in dry forage—more in some areas if the nutrition level in the feed is low.

Most of the ranches discussed here are owned by families, but some are part of large corporate operations. Clayton Williams, Jr., for example, operates ranches that he owns and others that he leases. The O'Connor Ranches are all owned by members of the family and operated in conjunction with each other by the same crews of cowboys. The Brown Ranch is run by the third generation of the family. The 101 Ranch is a smaller ranch operated by a family with little outside help. The owner of Nail Ranch died and left it in a bank trust so that his grandchildren can operate it later. A trust officer oversees the business operation, and a ranch manager sees to the day-to-day operation. The arrangements vary widely, but the purpose is the same—producing quality livestock for market, and at a profit.

Ranching is as varied as any other business, one that requires large investments of capital and time. *Economics* is the watchword, and profit and loss determine whether the operation can continue. The return on an investment of this sort is less than that on a reliable "blue chip" stock and is risky. The romance of ranching disappears quickly in the face of financial reality.

Drought is always a consideration. As I began working on this material in 1994–1996, drought gripped Texas, especially West Texas. Ranchers dumped their cattle on the market, prices plummeted from oversupply in the marketplace, and feed prices soared, the result of more dry weather in grain-producing areas.

By the spring of 1997 ample rainfall had returned, and ranges produced abundant forage for stock. However, the severe winter of 1996–1997 had created disastrous losses for ranchers on the Northern Plains, losses in the Texas Panhandle numbered in the thousands, and the pastures in Mexico were bare of breeding stock. Fewer available cattle means more demand, and prices rose to meet this market situation. Those with cattle prospered at the expense of those who had suffered the loss. By the summer of 1998, yet another weather crisis hit ranching with the hottest summer then on record. Following a wet winter triggered by El Niño, the horrible heat and drought that struck Texas cost the state an estimated $1.7 billion in agricultural losses. By the end of 1999, and throughout the summer of 2000, the situation worsened. Ranchers have to exercise creativity or go broke during these times of challenge.

Oil has long been important to the success of ranching in Texas. Albany rancher Bob Green said of the importance of oil, "A rancher needs the oil income to keep up his fences and keep his pastures in shape." Millions of dollars of income from oil have offset ranchers' losses on livestock and other agricultural efforts. In fact, oil has driven the ranching business, and many oil people have invested heavily in ranching. Not all ranches have the support of oil. For example, the Long X has no oil, nor does the 101 or Hudspeth. During the late 1990s the price of oil was typically at or below twelve dollars a barrel, well under production costs. By early 1999, it had dropped to less than ten dollars a barrel. As an Abilene geologist observed, there is no longer "an oil business." By 1999, oil prices had rebounded to more than $20.00 a barrel, and in 2000 the price has continued to climb.

Whether with oil or without, ranchers look for ways to generate capital. One method gaining popularity is the leasing of hunting rights for deer, turkey, quail, wild hogs, and other game animals. City dwellers who want a place to go in the fall are willing to pay for the privilege. Abandoned bunkhouses now shelter hunters. Even though having hunters on the land constitutes a liability, ranchers need the income from this activity. Fees paid for other recreational uses—such as fishing or "dude-ranch" arrangements that allow city dwellers to work alongside cowboys—help ranchers through these crises.

The following material is presented not as a scholarly exercise in comparative study but instead with the intent of telling the stories of these ranches, all historic but in different ways. The scholar may be disappointed; the general reader and the interested admirer of the way of life will find much to entertain and enlighten. I have depicted realistically what I saw on ranches when I visited.

The terminology used to describe activities and equipment is often unique to ranching. The Glossary will provide help to those seeking to understand this world.

From the Irwin-Clayton J Lazy C Ranch near the Clear Fork of the Brazos River, deep in ranching and cowboy country.

South Texas

SOUTH TEXAS IS THE BIRTHPLACE OF OPEN-RANGE RANCHING in the United States, at least according to Walter Prescott Webb, whose *The Great Plains* is one of the early studies of ranching in Texas. The diamond-shaped area he described, with San Antonio on the north, Indianola on the east, Laredo on the west, and Brownsville on the south, was certainly one of the first to be ranched. By 1600 Mexican vaqueros, mounted on mustang horses descended from those brought by the Spanish, were already working large herds of cattle on the open plains without the aid of pens or other facilities. Today, this area makes up the Brasada, the brush country that J. Frank Dobie, a noted Texas folklorist, made famous in his work, especially in *A Vaquero of the Brush Country*.

In time of drought, which frequently strikes this part of the state despite its proximity to the Gulf of Mexico and the Rio Grande, the area is characterized by stark barrenness. In wet years, however, the desert blooms with rich grasses, dappled by weeds topped with colorful flowers. Various species of thorny brush and cacti cover the sandy soil with a blanket of green. The winters are often mild this far south, and the summers hot and humid. It is a demanding landscape, but one on which cattle thrive, especially Brahman cattle, which seem more resistant to the heat, humidity, parasites, and disease that threaten livestock here more so than in other areas. When Col. Robert E. Lee, later commander of Southern forces during the Civil War, rode across the region in the 1850s as an officer in the U.S. Army, he saw ranching potential, as did men like Richard King, founder of the famous King Ranch, and Mifflin Kenedy, King's partner in both ranching and steamboat ventures in early Texas. The area was called the Wild Horse Desert because of the vast herds of feral mustangs that roamed there. These herds, bred over decades from stock escaped from Spanish explorers and early travelers, furnished the horses for the early development of Texas' ranching industry. Here, also, herds of feral Longhorn cattle, descended from animals explorers brought to the New World, provided the initial breeding stock for cattle ranching.

When Europeans first arrived, much of the land was open plain dot-ted with mottes of trees, but the heavy brush encroached steadily northward until it covered the plains to the Hill Country north of San Antonio. Today the brush forms a seemingly impenetrable wall of thorny growth that tests the mettle of man and beast. Composed of mesquite, prickly pear, huajillo, retama, granjeno, agrito, white brush, black brush, tasajillo, lechuguilla, and other thorny plants, this brush nonetheless provides very nourishing browse for livestock and other animals. As H. D. House, a South Texas ranchman said, "If I clear my brush, I have destroyed what my cattle eat."

One of the unique features of the area is the tall clumps of prickly pear, a cactus with large, rounded, fleshy pads covered with both fine, short and large, long spines. When forage is scarce during the winter or times of drought, ranchers burn the spines off with propane torches to provide feed. In no other part of the state is this practice common.

The region's abundant wildlife includes whitetail deer, bobwhite and blue quail, doves, javelina, and antelope. Bobcats and coyotes are common, and cougars also inhabit the region. One unusual species is the caracara, a black vulture with white markings that has extended its range north from Mexico but is not yet widespread in Texas.

Hunting in the Brasada demands creativity. The most common method of hunting deer is to cut *senderas*, paths through the brush, and watch for deer to cross or travel down them. But Graves Peeler, the man largely responsible for saving the Longhorns from extinction in the 1930s, was an outstanding hunter who shot deer from the back of a running horse.

Because of the warm climate, especially the mild winters, the brush *jacale,* or hut, was sufficient shelter for early inhabitants. Their water came from streams, but well water, much of it heavily mineralized, became available once drilling equipment came into use. In some parts, artesian wells also provide water for livestock.

The area is not uniform in flora or climate, however. In *Cryin' for Daylight*, Louise Stoner O'Connor describes Southeast Texas as being very different from the Brasada. Early settlers, she says, found little to tame in the region of marshlands and grasslands of the Coastal Prairies. Yet on the westward portions of O'Connor properties, one finds the brush.

South Texas has several historic sites, including the Alamo (in San Antonio), Goliad, San Jacinto, and other places that figured prominently in the war against Mexico for independence. The brush-choked Nueces Strip, between the Nueces River and the Rio Grande, was once claimed by both Mexico and the United States, but controlled by neither, and it became the home of desperados who contributed to another colorful chapter in Texas history.

The strong influence of Mexican culture on South Texas ranching is obvious in the language, the foods, and the gear. The very name, Brasada, reminds us of the formative influence of the Spanish culture in Texas.

The famous ranches of South Texas include the King Ranch and Punta Del Monte (both depicted in *Historic Ranches of Texas*), as well as the Alta Vista, Canales, Catarina, O'Connor, and Ray ranches, all large and important, but each unique in its own right.

Alta Vista Ranch

THE JONES FAMILY HAS BEEN A QUIET BUT VITAL FORCE in business and ranching in South Texas for at least five generations. Their ranch properties lie in Jim Hogg, Brooks, Starr, Hidalgo, and Duval Counties. Their 44,500-acre Alta Vista Ranch sprawls across a portion of the Coastal Plain of South Texas. The headquarters sits some thirty miles south of Hebbronville in Jim Hogg County.

The Jones family came to Texas in 1826, and later members of the family figured prominently in life around Beeville, Corpus Christi, and Hebbronville. Capt. Allen Carter (A. C.) Jones was born in Nacogdoches County in 1830 in what was then Mexican Texas. Although he lacked the opportunity for formal education because of circumstances of the day, Jones read widely and achieved a remarkable degree of literacy.

Jones matured quickly in the demanding region, where settlers often encountered trouble. His toughening for battle came in skirmishes with Indians and bands of raiding Mexicans in the Nueces Strip. Between the Rio Grande, which the Texans claimed as the border with Mexico, and the Nueces River, the border claimed by Mexico, was a no-man's-land where no law was respected or obeyed except that of the quick gun. Jones handled himself so well that he was elected sheriff of Goliad County before he turned twenty-one.

When the Civil War broke out, Jones enlisted as a private in Waller's Battalion under the command of Gen. Dick Taylor. After eighteen months in service, he was promoted to the rank of captain and sent to West Texas to serve under Col. Rip Ford and Capt. Santos Benavides, both legendary leaders for the Southern cause. After being wounded by a shotgun blast to the face when mistaken for a renegade, Jones recovered and went on to lead his men well. At the end of the war, he led them in retreat before a superior Union force that had landed at Brownsville, but Jones and the others in the unit, led by Col. Ford, fought a delaying action at Palo Alto, the site of a major battle during the Mexican War. This engagement was the last one of the Civil War and actually occurred after the war had officially ended. Jones then led his men to Beeville, sold their equipment, disbanded the unit, and retired from military duty. His company never surrendered.

In Beeville, Jones became a merchant and married Jane Fields; his first wife, Margaret Whitby, mother of his children, had died in 1854. He also became a banker. He retired from his mercantile business in 1884 but continued to serve South Texas. Considered the father of Beeville, Jones was instrumental in bringing two railroads to the town—the Aransas Pass Railway and the Gulf, West Texas, and Pacific Railway. He also helped found the First National Bank and was general manager of Beeville's oil mill. Not just a businessman, Jones was also a cattleman and oversaw two thousand acres of farming. His death in 1905 brought an outpouring of recognition for Jones' contributions to the region's development.

His son, W. W. (Bill) Jones, was a powerful figure as well. Born in 1858 in Goliad, he learned early to love life on the range and decided to be a cattleman, despite his father's wishes that he become a merchant. He helped trail cattle from the Coastal Plain to Kansas during the trail drive era and later was a charter member of the Texas Cattle Raisers Association, a forerunner of the Texas and Southwestern Cattle Raisers Association, still a vital force in the cattle business in Texas. He hosted the association's meeting in Corpus Christi in 1931. During his years of study at Roanoke College in Salem, Virginia, and Poughkeepsie Business College in New York, Jones developed the family motto, which he displayed on a placard in the room where he studied—Labor Always Wins.

Although he preferred ranching to commerce, Bill did fulfill some of his father's expectations. He founded the Alice State Bank and Trust and Corpus Christi National Bank (now Bank of America Corpus Christi). After the devastating hurricane of 1919, he took over and restored the Nueces Hotel, one of the few structures that survived in downtown Corpus Christi. The landmark building near the waterfront is still standing. He also helped in constructing the Nueces County Municipal Building. His dream of making Corpus Christi a deep water port was aided by Robert Driscoll, Jr., Richard King III, O. W. Keller, and others who created the Nueces Navigation District and finally saw the port opened in 1927.

Bill Jones was primarily a cattleman, one known by the white Panama hat he wore in the summer and the high-crowned, broad-brimmed black felt he preferred for winter. By 1890 he was buying ranch land in McMullen County, and he later bought a large part of what is today Jim Hogg County. Some of the land he purchased was already organized into ranches, but some was just raw land. One of his early acquisitions came when he lent money to the owner of a ranch who became unable to make his loan payments. He asked Jones to take over the ranch and cancel the debt, which he did. Jones then kept buying land until his holdings stretched sixty miles north and south and four miles east and west to include some 300,000 acres. He later organized Jim Hogg County, and his son A. C., his grandfather's namesake, became the county's first judge.

Jones' favorite ranch was Alta Vista. Purchased between 1888 and 1890, the ranch originally had a town site with a post office, a store, and even a dance hall. The idea of a town did not catch on, but the ranch is still the headquarters for the Jones family's

several ranches and the home for the foreman and twelve to fifteen vaqueros who work on them. For efficient operation, ten pastures and as many smaller traps to hold cattle divide the ranch into working units. Three strategically placed sets of corrals provide facilities to work the cattle.

All of the family ranches primarily stock cows and bulls, called a cow-calf operation. The initial stock for these was Santa Gertrudis from King Ranch, but later Jones turned to Hereford cattle. The breeding pattern lately has been to crossbreed Herefords, particularly with the Brahman that has proved so durable in the region's heat and humidity. Most of the cattle on the ranch can be said to be of mixed blood, but they are horned stock, indicating some Hereford influence. All of the cattle have Brahman traits, including the widely admired tiger-striped coloring and characteristic vigor and spirit.

The Alta Vista usually has two roundups per year. The first of these, for branding, vaccinating, and otherwise working the calves, takes place in April and May. In the fall, roundups for shipping usually begin in October and end by December. Originally cattle were trailed to the railroad line in Hebbronville for shipping. Now, of course, the cattle are hauled in cattle vans to market.

Gathering the cattle in the heavy brush is still occasionally done by vaqueros on horseback, and sometimes dogs are used, but the ranch often resorts to using two or more helicopters to assist in the gathering. These small two-man craft dart low over the brush to drive the cattle out. Some cattle are pasture-wise enough to escape into thick brush and lie down to avoid detection.

The horses used by the vaqueros grow up on the ranch. The Jones family has never been active in showing or racing horses, or developing cutting horses for competition. The purpose of the *remuda*, or horse herd, on Alta Vista is providing suitable animals for working in the heavy brush. This is not country where riding is done for pleasure. In fact, since the dangerous thorns can cripple a horse, it makes no sense to ride prized horses to chase cattle. Riders must crouch low to the necks of their mounts to avoid being swept off by low-hanging limbs, and they need horses wise to the brush and the ways of cattle. Both geldings and mares are broken to ride, and the ranch keeps a string of three horses per vaquero.

Only high-quality tack can withstand the demanding work. The saddles used are typically styled after those made famous by the King Ranch Saddlery in Kingsville. They feature swelled forks for protecting the legs and rolled cantles. This type of saddle, used by ranch hands throughout Texas, descended from the traditional Texas double-rigged saddle, which has two girths. The bits used on young horses are snaffles, which are hinged in the middle, but the shallow port bit is commonly used on older horses. Unlike the large Chihuahua spurs found in Mexico, the spurs used here have small rowels. To protect riders' feet from the thorns and to keep their feet from being knocked out of the stirrups by the brush, toe fenders, or *tapaderos*, are used. Heavy leather chaps and tough canvas brush jackets are essential to surviving in the Brasada. Ropes are thirty feet long, as they are in West Texas and the Panhandle, and roping methods include both dallying, wrapping the rope around the saddle horn, and tying it, which is called hard-and-fast.

One technique that distinguishes South Texas cowboys from those in other regions is the use of a single rein tied at each end to the bit and looped over the horse's neck. Cowboys elsewhere prefer split reins that are not tied together.

Evidence of a strong vaquero tradition on the ranch is easy to find. Frank Graham, long-time foreman for the Jones family, spoke Spanish as fluently as he did English. In fact, the working dialect is an interesting mixture of both languages. The food on the ranch is often in the Mexican tradition, including flour tortillas, *carne guisada* (a kind of Mexican stew), and small tacos, or taquitos. Some of the men

LEFT: *Ranch manager Harold Lee Henry*
RIGHT: *W. W. Jones III*

also engage in traditional crafts such as twisting horse hair into ropes, though use of plaited rawhide is not common.

The ranch is fenced primarily with barbed wire on cedar posts, but some metal posts are being used in new construction. There is very little net wire because sheep raising, once popular in the region, fell out of favor after ranchers continually suffered losses to predators, especially coyotes. Net wire is still seen occasionally on pens, but ranchers prefer to use metal panels on new or remodeled pens.

Hunting is important to the Alta Vista's financial success. The ranch leases its property to individuals who organize and manage the hunters so that ranch personnel are not involved. The area's abundant game attracted the late Sam Walton, founder of the Wal-Mart chain, who once brought as his guest former president Jimmy Carter.

The thorny brush is a constant factor in the lives of anyone who lives or works in this region. Although efforts to control it have made parts of the ranch open, productive grassland, the brush continues to encroach on and choke out the more desirable grasses. As already mentioned, however, the brush is not altogether harmful, for it offers browse for the livestock and also food and cover for the wildlife on which the ranch's hunting enterprise depends.

Providing adequate water for livestock is a constant concern. Stock water is pumped from wells by windmills. Some of the twenty-three windmills on the Jones' ranches are constructed of lumber, but the more recent ones are on steel towers. Most of the windmills have storage tanks to supply water even when there is no wind, and the water is fed into a trough by a valve controlled by a float. One of the daily chores is checking and servicing the windmills and making sure that the water is available to the stock.

The Alta Vista's business office is in the Bank of America building on Water Street in downtown Corpus Christi, the closest major city. W. W. (Bill) Jones II ran the ranch for much of his life, from the time he returned home from military service in World War II until his death in 1998. He learned his work from his father, A. C., who, he recalled, never told him to "go" do something but instead said "follow me." Careful management of the ranch is essential since it depends primarily on cattle and has no oil deposits to provide a financial safety net. Recalling his experiences running the family operation, Bill says, "I've walked over those ranches in heat and cold, rain and shine, and I know every foot of the country. I was always impressed by my mother's telling me that the best impression I could leave on the land was my own footprints. I've also always tried to keep to the family motto, 'Labor Always Wins.' "

Following in Jones' footsteps is his son A. C. (Dick) Jones IV, who, though he keeps an office in Corpus Christi, is often found rushing across the thorny range in his Chevrolet Suburban or on horseback. A man born and bred in the brush, he is as much at home there as he is in his office.

In 1996 Dick Jones formed an operating company, Jones Ranch LLC, which currently manages the Alta Vista, Alto Colorado, and Borregos Ranches, a large part of the original estate. He is assisted by his son, W. W. Jones III, and foreman Harold Lee Henry, the son of Doc Henry, long-time foreman for Kathleen Jones Alexander, Lorine Jones Lewis, and Alice Jones Eshleman, all daughters of W. W. Jones. Lorine, who inherited the Alta Vista from her father, left the ranch to her nephew W. W. Jones I. After his death in December 1998, the ranch was passed on to his four daughters, Susan Jones Miller, Lorine Jones Booth, Kathleen Jones Avery, and Elizabeth Jones.

The Alta Vista Ranch is a fine example of the combined influence of Anglo and Mexican heritage. The traditional life on the ranch, embodied in the vaqueros, is testament to the validity of the methods developed early by ranches in the area and still considered sound. The Jones family continues to be a dynamic force in ranching and business life in South Texas.

Canales Ranch

THE CANALES RANCH IS UNIQUE AMONG THOSE INCLUDED in this volume because the Canales name traces back to the time when Spain controlled the area that is today Texas. In fact, Don José Canales was among the first settlers to enter Spanish Texas. Whether he was a direct ancestor of the founders of Canales Ranch is unknown. Although some settlers had migrated in the first half of the 1700s to what is today South Texas and the area adjacent to it south of the Rio Grande, in 1748 the Spaniards mounted a major effort to colonize that part of their territory. Don José de Escandón, a native of the province of Santander in Spain, was authorized to take a group of settlers to the area with the intention not of military conquest but, instead, of establishing families on ranches to deter encroachment by other European powers, especially France.

Escandón eventually settled about 9,000 people, mostly from Querétaro, Nuevo León, and Coahuila, Mexico. Only about 150 of these were soldiers. The others were ranchers and their families, people accustomed to privation and hardship on the frontier and familiar with stock raising. They brought their belongings in wagons and drove their livestock, mainly cattle and horses. They also found feral cattle and horses here in large numbers.

The people came to make a life for themselves in a new land, not to establish missions to convert the Indians to Catholicism or to make slaves of them (Graham, "Spanish and Mexican Origins," 81). Among the villages they established are Camargo, Reynosa, Mier, and Revilla in Mexico. Reynosa was founded by fifty families composed of 238 people. Only 11 soldiers were among them, so this was no fort. Mier and Revilla were established as ranch headquarters, as was Matamoros, the principal Mexican city in the region today. Originally the site of a *rancho* known as San Juan de Los Esteros, the settlement would later be called Congregación del Refugio and then Puerto del Refugio. In 1826 it was formally organized and named after Mariano Matamoros, a priest and soldier martyred in the battle for independence from Spain. Escandón selected the names for some of the villages he established from those of his native Santander, and the region on both sides of the river was known as the province of Nuevo Santander. The Mexican portions became part of the state of Tamaulipas.

Across the river, in what is today Texas, the pattern was much the same. Laredo, established in 1755, has a history similar to that of Matamoros. It was founded by Tomás Sánchez de la Barrera y Garza and three other families from Dolores, Mexico. The original community was principally devoted to raising livestock. Rio Grande City began as a ranch headquarters known as Rancho Carnestolendas. Roma was originally Rancho Buena Vista. Not all of the ranches became towns. The Santa Anita grant of Manuel Gómez, for example, became the ranches of the families of John McAllen and John Young in Hidalgo County (Graham, "Spanish and Mexican Origins," 82).

In 1757, a royal commission from the King of Spain was sent to the area on a *visita general*, or general visit, and assigned to each settler a *porción*, a portion of land that fronted the Rio Grande for 1,300 varas and extended back from the river 2,500 varas. (A vara is a Texas unit of length equal to 33.33 inches.) The purpose was to guarantee each settler access to the reliable waters of the river, a necessity in this typically arid region. By 1760 more than 350 *rancherías*, or ranch headquarters, existed in the area with more than three million head of livestock. Because of the region's remoteness and the absence of nearby markets for beef, ranchers drove herds of cattle to sale in Louisiana and into Mexico.

As a prosperous ranching culture evolved in the region, raids by Indians intent on driving out the new settlers began to escalate. When the revolution for Mexican independence from Spain drew soldiers away and left the settlers vulnerable, many of them fled the region. After the war between Mexico and the United States ended in 1848 with the Treaty of Guadalupe Hidalgo, Anglo settlers moved into the region from the east to establish their own ranches on the abandoned land with herds of feral livestock. Among those were Richard King and Mifflin Kenedy, former steamboat captains on the Rio Grande during the Civil War and the Mexican War. They bought land grants from the Mexican owners who chose not to remain in the region, but not all of the Mexicans fled. Some remained to carve out ranches of their own.

The Canales family comes from this frontier background. On the Canales side it goes back to the mid-1800s, some five generations. On the maternal side is the name Cavazos, which goes back nine generations. In 1876 Andrés Canales married Tomasa Cavazos and joined the two families. Tomasa was a descendant of Capt. Don Blas María de la Garza Falcón, the original military commander and leader of the town of Camargo in Mexico. He and his wife, María Josefa de los Santos Coy, had a daughter, María Gertrudis de la Garza Falcón, who married Don José Salvador de la Garza, an original settler of Camargo and the recipient of the large El Espíritu Santo land grant from the King of Spain in 1781. It originally contained 59 *sitios*, or square leagues, of land, more than 262,000 acres. Today, the city of Brownsville, Texas, the seat of Cameron County, sits on the tract.

The Canales family still has a document describing the land as bounded on the west by a "dense

CANALES RANCH 25

Gus Canales feeding horses

thicket or woods extending from the Arroyo Colorado on the north to the Rio Grande on the south and passing through a place called Puerta Vieja." The Arroyo Colorado is a stream flowing from Lake Llano Grande in Hidalgo County to the Laguna Madre on the coast of the Gulf of Mexico. On June 20, 1834, the governor of the state of Tamaulipas, Mexico, confirmed the grant on the heirs of Salvador de la Garza. By this time, Mexico had obtained independence from Spain. The confirmation of the grant was requested by Juan José Tijerina on behalf of his wife, Feliciana Goceaschochea, one of the heirs of José Salvador de la Garza.

There was a strong military presence in the family, including the husband of Francisca Xavier de la Garza, a man named Don José de Goceaschochea, who was a captain in the army and later a chief justice. His daughter Estefana married Francisco Cavazos from the Reynosa area, and later married Trinidad Cortina, with whom she had three children. One of them was Juan Nepomuceno Cortina, later Gen. J. N. Cortina, a Mexican folk hero who started two "wars." In 1859 he invaded Brownsville, Texas, and took control of the town to avenge a wrong to one of his employees. In 1861, as a Union sympathizer, he invaded Carrizo, Texas, the seat of Zapata County. Cortina drew a lot of attention on the border during his career, during which he was accused of heading a ring of cattle thieves operating in the Nueces Strip, that disputed no-man's-land between Texas and Mexico. That he was a strong supporter of Mexican citizens in Texas and a prominent figure in the area cannot be denied.

The Cavazos family joined with the Canales family through a descendant of Feliciana Goceaschochea and Juan José Tijerina who married Tomás Cavazos. In 1876 their child Tomasa Cavazos married Andrés Canales, whose ancestor Servando Canales was a general in the Mexican army. Andrés Canales had come to South Texas from Mexico in 1860 at age eight with his father Albino and mother Josefa González de Canales from the area near Mier and lived on Los Tramojos Ranch. Later, he and his wife lived on Los Veladeros Ranch, owned by Sabas Cavazos, her grandfather. It was located about twenty-two miles from the Santa Gertrudis Ranch of Richard King. After he incurred heavy debt from financing a failed revolution in Mexico, Sabas Cavazos sold part of his ranch to Capt. Richard King. It became part of King Ranch.

In 1878 Andrés Canales established La Cabra Ranch northeast of Premont in what is today southwestern Jim Wells County. He first devoted his efforts to raising sheep, but he soon began raising fine cattle and horses. To pasture the animals, he purchased other land, including a 12,000-acre tract owned by the state of Texas. He also purchased two other ranches in Jim Hogg and Brooks Counties totaling approximately 15,000 acres. In 1892 he stocked the ranch with 2,500 head of cattle, and he became widely known as a breeder of excellent cattle and horses. To promote settlement of the region, Canales donated land for the railroad right-of-way, and in 1904, when Jim Wells County was formed, he saw the town of Premont founded on land that he had owned. He was also influential in founding the Bank of Premont.

The oldest son of Andrés and Tomasa was José Tomás Canales, a cofounder of LULAC, the League of United Latin American Citizens. Mr. Canales was an attorney and also served as a representative for Cameron, Zapata, Starr, and Hidalgo Counties in the Texas legislature and as a county judge. As a state representative, he authored the Canales Act, the bill responsible for downsizing the Texas Rangers to a small, elite group of lawmen.

A reminder of the historic past of the Canales Ranch is a jacale on the southern part of the property. It stands by a small lake in an area that for decades served as a gathering ground for herds before the ranch was cross-fenced into pastures. The cowboys would spend days rounding up the cattle on various parts of the ranch and driving them to this watering hole. The cattle to be marketed were then herded to the north end of the ranch and on to market. The cattle left on the ranch were allowed to graze where they pleased. Later, after the railroads were built, the ranch hands drove cattle to the large complex of railroad shipping pens at Hebbronville, some thirty-five miles to the east. Now the cattle are transported in large vans.

Management of the ranch today involves complex arrangements and leases of property inherited by various members of the family over the last several decades. These properties lie in Zapata, Brooks, Jim Wells, Jim Hogg, Kleberg, McMillen, LaSalle, Webb, and Duval Counties. Gus T. Canales, an eighth-generation Texan in the family line of Cavazos and a fourth generation of the Canales family, manages much of the property for family members who live and have careers far from the remote stretches of brushy landscape around Premont. There are so many heirs that it would be impossible for all of them to live on ranch property even if they wanted to. And not everyone has the interest or time to devote themselves to the demands of ranching, or subject themselves to the heat, cold, dust, and uncertainty. One heir, however, Marc Cisneros, a retired general and president of Texas A&M University, Kingsville, still has partial management responsibility for his great-grandfather's Los Tramojos Ranch.

In the early 1900s the Canales Ranch herds included Hereford cows and often Brahman bulls. By the 1940s some Zebu bulls were being bred to the Herefords. Later the family decided to cross Brahman cows, which fared well in the hot, humid conditions, with Hereford bulls. This allowed the ranch personnel to focus most of their care on the smaller number of bulls because the hardier Brahmans required much less attention. Later the ranch stocked Santa Gertrudis cattle and then, recently, Santa Cruz cattle, two breeds developed by King Ranch. The ranch has also used Romanola bulls, a European breed resembling a Brahman but with no hump and smaller, rounded ears. This breed seems well adapted to the region. All of the Canales cattle bear the family's C brand, which has been used for several generations.

Although most of the family's efforts have been devoted to cattle, attention to raising horses has also been keen. In the ranch's early history, A. T. Canales, son of Andrés, bred horses that were mainly utility animals used to work cattle, called in Spanish *caballos trabajo*, or horses for work. Even at that time, however, the owners and ranch hands were interested in horses for other reasons. These *caballos finos*, or fine horses, were usually well bred, fast, and high spirited.

Today the ranch hands ride geldings, mares, and even the stallions when working cattle. Gus T. Canales seeks a "good, solid, honest cow horse" and breeds for a medium-size horse with a good disposition and lots of cow sense. He is currently using a line-back dun stallion he acquired from Frank Graham, a prominent South Texas cowboy and horseman. For many years Graham worked for the Jones family, whose ranches are also included in this volume. Graham has long taken pride in breeding these dun horses descended from Spanish stock brought to North America centuries ago.

The tack and roping gear used on the Canales Ranch are typical of this area. The older saddles were made by Frank Vela from Floresville and have tall swells in front and straight cantles. But most of the young men ride saddles they have won at ropings, rodeos, and other competitions. These are swelled-fork, low-cantle, double-rigged saddles. Gus T. himself rides a saddle made by Shirley Brown of San Antonio. It is a heavy one with high fork and cantle to protect the rider in the brush. The bits are low port, and the reins looped. Gus T.'s favorite rein is a looped or "roping" rein introduced to him by one of the old foremen, who taught him to tie a single leather string to the rein at its midpoint and to hold that string when riding. He even ties his horse by this string, which breaks easily if the horse pulls back while tied. This practice saves breaking a rein or an expensive headstall. As on other South Texas ranches, the lariats are about thirty feet long.

Water for the stock comes mostly from wells with windmills and electric pumps pulling the precious liquid to the surface. The water is pumped into large storage tanks and fed into troughs for the stock. The ranch also has plans to test solar-powered pumps.

Fencing is mostly of the tall net wire developed years ago by King Ranch, and pens are built with welded pipe and wooden planks. Canales says he plans to depend more on wire and less on wood and pipe because it makes for pens that are cooler on the stock and less expensive to build. Most of the old *corrales de lena* made of stacked mesquite logs have rotted and disappeared or burned in range fires.

Brush management has been a long-standing problem on the ranch. The tall clumps of prickly pear common to this area are both a blessing and a curse. Cattle eat it, but eventually they develop health problems and can even die from ingesting the thorns and the fiber that make up the leaf structure. Controlled burns are proving useful for controlling the brush in years when rainfall is plentiful and enough plants grow to fuel the fire. Removing brush with a bulldozer is also common, but the brush regrows with considerable vigor, requiring additional attention in the form of grubbing or chemical spraying to control it. When the pastures are burned and the season is wet, a fairly rare occurrence, the resulting growth is fresh and very attractive to livestock and wildlife. Hunting of deer and other game is a popular sport for family members.

Gus T. Canales has a positive outlook on ranching. He says, "I feel fortunate to have ancestors that accumulated and kept what they owned and passed it on so that I can still know some of what they did. I also feel grateful that I have relatives who have entrusted me to manage the land for them. I think that if we do not continue to have that trust in the manager, the honest and caring steward of that property, it will start breaking up, and once you start getting small property owners, it will dissolve. The on-site manager must have affection for the land, and likely will be a member of the family. An outsider just will not understand the ranch and the traditions on that ranch. But it could be acquired, *if* they lived on it long enough."

Dawn Canales, Gus T.'s daughter, expresses the feeling of the younger generation for the land: "To me South Texas is many things. It not only signifies home and family but also a deep feeling in my heart that is difficult to express. My father tells me he remembers when he was young and the ranches were open country like prairie land with few brush mottes, unlike the brushy land we constantly try to manage today. We as humans alter our environment tremendously. I can only hope and pray that our generation and following generations will be wise and strive to keep South Texas as native as possible; after all South Texas is also known as God's country."

The Canales operation owes its past to the rich Spanish-Mexican culture that came to South Texas and established ranching as a tradition on a raw frontier. Here, from the Mexican mounted herder, the vaquero, developed the Texas cowboy whose influence—and image—have spread around the world.

The many individuals who belong to the extended Canales family have a proud heritage that includes ranching, politics, military service, medicine, law, and other professions. This lineage goes back to early days of European settlement in Texas, especially in the Cavazos and Canales names. Today, the ninth generation of this family continues to carry on a tradition in South Texas, where ranching will long be a vital way of life.

Catarina Ranch

CATARINA RANCH SPREADS OVER 100,000 ACRES IN DIMMITT County of South Texas and is about twenty miles southeast of the town of Carrizo Springs. The ranch is owned, as it has been for decades, by the family of former governor of Texas Dolph Briscoe, Jr. This cow-calf operation has been modernized in every way possible, but it clings fast to a colorful ranching tradition. The acres held by the Briscoe family are but a portion of the original Catarina Ranch, and the story of the development of the ranch and the region is an important part of Texas history.

Early settlement in this area followed a fairly typical pattern. The initial settlers came in oxcarts from East Texas and Mexico, and brought with them herds of Longhorn cattle and flocks of sheep to graze on the excellent grasses found growing in abundance. This part of the state is also blessed with water from many flowing springs at which early settlers watered their stock and which give the town of Carrizo Springs its name.

The devastation of the 1886–1888 drought wrought havoc here, as it did throughout the West. Many animals died of starvation, and because the range was overgrazed, brush, which had previously been controlled by fires set by Indians and others, began to encroach, eventually spreading across the ranch land. Coyotes took advantage of the new cover to prey on sheep, causing ranchers to give up on raising sheep in this part of Texas by 1900.

Among the early settlers were Mexican vaqueros who came to be part of the ranching culture. Some worked as ranch hands, but others owned their own land and cattle. These early settlers lived in jacales, brush huts which served as adequate shelter from the heat and other threatening elements. Mild winters made such living quarters habitable year-round.

One way for the early settlers to raise money was "mustanging." Those who were bold, brave, and able would capture wild horses from the vast herds either by wearing them down by following them on a series of grain-fed horses, or by trapping them in makeshift corrals. These horses were notorious for being spirited, but capable riders could bring one to accept the saddle or plow. A common practice of mustangers was to capture several horses and ride them just enough to get them to accept a saddle. Then the men would drive the "green-broke" horses to towns where they could sell them for prices high enough to make the effort worthwhile. No record exists, however, of anyone getting rich by breaking and selling mustangs. In fact, the demanding and dangerous work was often done as a last resort for raising capital.

Because of the area's shallow water table and farming potential, it was once part of the fertile winter garden lying along the Rio Grande to the south of Carrizo Springs. After developers brought in equipment, drilled wells, and plowed the land, farming became the order of the day. Initially, onions, spinach, and strawberries were grown, and later oranges and grapefruits were introduced. Ultimately, though, farming declined, in part because of the economic troubles of the Great Depression, but also because the underground aquifer was inadequate to support extensive irrigation.

Thus, the land was returned to grazing. In the 1860s and 1870s, herds of cattle from the area were driven to market, some up the Chisholm Trail and others by way of the later Western Trail. The railroad, of course, eliminated the need for long-distance trail drives. The International and Great Northern Railroad also eliminated the need to haul supplies by wagon from San Antonio, the nearest town to the east, or from Presidio, the closest town of any size, on the Mexican border to the west.

Open-range grazing continued for some time, but the introduction of fencing eventually made that style of ranching impossible. Those who did not own land had to push on to other areas, but those who stayed developed a more modern version of ranching.

Catarina Ranch lies in the Nueces Strip, that hotly contested area between the Nueces River and the Rio Grande once claimed by both Texas and Mexico but controlled by neither. Mexican banditos, American outlaws, and Comanche, Kiowa, and Apache Indians were known to frequent the area. The herds of wild horses in the Strip were important to these mounted bands, and conflicts between the groups were bitter and bloody.

During the Civil War, Texas Rangers and the Frontier Regiment were the only armed forces that kept the Indians, outlaws, and Mexican raiders at bay. These organizations had relatively few men, considering the vast area that they were expected to police. A Ranger force led by Capt. Mat Nolan patrolled the area and had at least one fight with a band of Mexicans at Carrizo Springs. After the war ended, some Anglos settled in the area. Family names included English, Burleson, and Vivion.

For years Indian raids around Carrizo Springs were common occurrences. In 1866, Ed English and some other men barely escaped an Indian attack in which English suffered an arrow wound. But the last raid was in 1870, when two hundred Indians attacked the Charles Vivion Ranch, killing one man and kidnapping a boy. This same band pushed on toward Carrizo Springs, but they were driven off in a desperate fight near the ranch of Dave Adams, who was also killed. Peace finally came to the area when a regiment of African American Buffalo Soldiers led by Lt. John Lapham Bullis exerted enough pressure

to thwart the raids. Nevertheless, horse and cattle thieving continued in the area until the turn of the century.

Catarina Ranch was founded in 1881 by David Sinton. The name *Catarina*, Spanish for *Catherine*, had been associated with the area for long before Sinton established the ranch. In the late 1700s a Spanish priest named Friar Morfi who was traveling on *El Camino Real*, the King's Highway, reported his location as being close to *Aguaje de Santa Catarina*, or the Waterhole of Saint Catherine. Various local stories of more recent events also offer explanations for the name's origin. According to one, a Hispanic woman by that name was killed here by Indians, and another says that Catarina was a woman who died while working for Charles Taft, a later owner of the ranch. Another involves Catarina Soto, a woman found dead on the ranch in 1913 and who may be buried there.

The Sinton family established several political ties for the ranch. David Sinton's daughter Ann married Charles P. Taft, brother of then-president of the United States William Howard Taft. The Tafts accumulated a considerable amount of real estate in Dimmitt County, including 235,000 acres that were Taft-Catarina Ranch, some land once owned by early Texas settler Wiley McGee, and some Spanish or Mexican land grant properties. All of this ranch land was managed by the Coleman Fulton Pasture Company and directly supervised by Joseph F. Green.

The Tafts' most significant and long-lasting contribution to the ranch was the construction of the Taft House, an impressive three-story structure designed to accommodate the president should he decide to visit. Laura Tidwell, in her book *Dimmitt County Mesquite Roots*, recounts how the house was constructed from materials hauled by wagon from Cotulla, more than twenty miles to the east. The first floor consisted of an entry hall with a powder room behind the stairs, living and dining rooms that shared a double fireplace, a pantry, and two other rooms with a bath, which together occupied a space measuring fifty-six by seventy-four feet. Also on the first floor were the kitchen, a storage area, and porches. The second floor had four bedrooms with fireplaces and two bathrooms. The third story was a "carefully finished" attic (112).

The rooms of the house were elaborately appointed. The walls featured wainscoting, and the ceilings were molded plaster. The floors were "geometric hardwood inlay, all hand-finished by the contractor's son, Joseph W. Brauer" (112). Tidwell notes that bronze nails were used in the construction of the building, and stories abound about the oversized bathtubs designed to accommodate the president's hearty girth. But despite its opulence, the house provided a residence for no one in its early days. Even its builder, the ranch manager Joseph Green, did not live there, but hired a caretaker to watch over the house.

S. W. Forrester, an oilman from Kansas, purchased the property from Taft in 1920, and his family was the first to actually live in the Taft House. Forrester searched diligently for oil in the region, especially on the ranch, but was unable to develop any significant wells.

The Coleman Fulton Pasture Company, which had leased the grazing rights, had to sell their cattle when Forrester sold the ranch to a syndicate. In perhaps the last cattle drive from Dimmitt County, Joseph Green, who by this time was manager of the Coleman Fulton Pasture Company, drove the cattle to another ranch near Taft, Texas.

The syndicate that purchased the ranch included C. H. "Clint" Kearney, J. F. Jarrett, and H. V. Wheeler, whose plans had nothing to do with cattle. In 1925 a massive effort began to subdivide Catarina Ranch and create farms, something that had already been done at the XIT and Renderbrook Ranches. The group hired Charles E. F. Ladd of Kansas City to supervise the project, and trainloads of potential buyers were brought to the region. After 1930, however, these efforts lost momentum because of the poor economy, lack of water, poor market access, and insufficient funds for the development.

At this point a strong force entered the picture: the Briscoe family. Dolph Briscoe had begun ranching in Dimmitt County in 1923 with Ross S. Sterling, president of Humble Oil and Refining Company and later governor of Texas. In the early 1930s he began leasing land on Catarina Ranch, and in 1938 he bought the 100,000 acres which has been in the family ever since.

Since his father's death, Dolph Briscoe, Jr., and his wife, Janey, have operated the ranch and used it as a retreat. Briscoe, a former governor of Texas, is an influential banker and stockman, and one of the largest landholders in the state. A native of Uvalde and a graduate of the University of Texas, he has served in the state legislature and has also been a director of the Texas Sheep and Goat Raisers Association, president of the Mohair Council, and president of the Texas and Southwestern Cattle Raisers Association, to name only a few of his important posts.

The Briscoe family focuses on a cow-calf operation based solidly on Santa Gertrudis cattle, a breed developed by King Ranch several decades ago. The breed's characteristics make it ideal for this hot, humid part of Texas. The ranch's office is in Uvalde, some fifty miles north of Carrizo Springs. Several of the ranch hands are Hispanic, so, understandably, Spanish is the working language.

The tack used is typical of other Texas ranches, including saddles with swelled forks and rolled cantles. Toe fenders, or tapaderos, are made from a heavy rubber belting material and provide protection against the thorns. When the men rope in the pasture, they are prone to dally rather than follow the early Texas custom to tie the lariat "hard and fast."

Three horses per man are adequate for the work on Catarina because the heavier roundup responsibility falls to helicopters, which are more efficient in the heavy brush. The ranch has not raised horses for several years because it is more efficient to buy them. The ranch no longer castrates the bull calves. Calves are simply branded and vaccinated and allowed to grow until marketing time, usually in the late summer.

Leased commercial hunting of deer, turkey, quail, and other game animals is an important part of the ranch's income. Hunting of deer and turkey is excellent, and hunting feral hogs in the off-season is also popular because these often fierce swine offer a unique challenge, especially for the archer or handgun hunter.

Winter forage for the cattle is supplemented when necessary by burning prickly pear, a time-honored practice in an area where the cactus is abundant and grass scarce. The prickly pear grows in large clumps, some reaching six feet in height. After the spines are burned off with a large propane torch, the protein-rich fleshy pads offer the domestic stock and wild game a succulent source of nourishment. Both will eat prickly pear thorns and all, but cattle can die as a result, hence the practice of "burning pear."

As do other ranches in this part of the state, the Catarina Ranch fights to control the spread of brush and keep their ranges productive. Root plowing and bulldozing, rather than chemicals, are the preferred means of controlling the problem. Getting rid of the tree trunks and limbs, most bristling with thorns, is accomplished by controlled burns.

The ranch uses both barbed and net wire, with increased use of net wire, and continues to rely on cedar posts, which, in this dry climate, last for many years.

One of the most significant recent developments at the Catarina Ranch came with the removal of the Taft House from its original location to Briscoe's ranch headquarters. In September 1981, Earl Bradford, an Austin-based specialist in moving buildings, cut the house into two pieces and moved it to its

COFFEE

current location. By 1994 the house had been restored to its previous splendor.

Catarina Ranch remains an active cattle ranch, with the dream of making it part of the agricultural winter garden only a memory. It is perhaps the best known of the several ranches owned by the Briscoe family, who rank sixth in total head of livestock for a cow-calf operation in the United States. An efficient ranch, the Catarina has played a unique role in the history of Texas ranching and continues to serve as a landmark for this proud tradition.

O'Connor Ranches

SPREAD OVER PARTS OF GOLIAD, REFUGIO, ARANSAS, AND Victoria Counties are the O'Connor Ranches. Established in 1836 by Thomas O'Connor, these ranches are still operated by the family and are evidence of the tenacity and vision of those who have played a prominent role in the diverse development of the area. Their accomplishments, especially those of the early generations, are described in a book rich in details about the history of the area, *Three Hundred Years in Victoria County*, by Roy Grimes.

This part of the state is in many ways ideal cattle country. In her book *Cryin' for Daylight*, Louise S. O'Connor, owner of part of the ranch properties, says that in Refugio, Victoria, and Goliad Counties is found a "geological pocket" characterized by "rich, black land high in potassium . . . and ample rainfall." These provide "an environment for natural grasses . . . extremely high in nutrients." She calls it a "gentle country" where there was "little to tame" when settlers first came. Water and food for both humans and animals are abundant, and the temperate climate stands in marked contrast to that further north and west (3).

The region has proved immensely productive for ranching since the early 1700s, when Spanish colonization efforts brought Catholic priests, soldiers, horses, and cattle to Presidio La Bahía, an outpost of the mission system. The presidio still stands near Goliad, twenty-five miles southwest of Victoria. In the 1820s, the Mexican government, hoping to encourage settlement of the area, allowed a group of Irish Catholics to organize a colony. A group of Irish Catholics already in Mexico were joined by newcomers from Ireland and together they established the Power and Hewetson Colony. Since these settlers had not been allowed by the British government to own property at home, one of their main interests in joining the new colony was possessing land (O'Connor, 5).

One of the Irish immigrants was young Thomas O'Connor, a most unusual and productive individual whose goal to own land was realized in impressive fashion. He came with his uncle, Col. James Power, a leader of the colony, and though only in his early teens, O'Connor received 4,428 acres of land in the Aransas-Papalote area. It was only a beginning for him.

When fighting broke out in the rebellion of Texas against Mexico, O'Connor answered the call to serve under Capt. Philip Dimmitt. He served with the Texas army at La Bahía and was one of the ninety-one signers of the Goliad Declaration of Independence, a document written by Dimmitt and Ira Ingram, who served as speaker of the house for the first congress of the Republic of Texas. O'Connor was the youngest Texan to fight in the Battle of San Jacinto on April 21, 1836, when Texas forces under Gen. Sam Houston defeated the forces of Gen. Antonio López de Santa Anna to end the war. In return for his army service, he received another 650 acres of land.

After the war ended, O'Connor built a log cabin along the San Antonio River near some relatives named O'Brien. His first wife, Mary Fagan, was from a neighboring family, and they were married in the San Fernando Church in San Antonio. She died in 1842. O'Connor was married again in 1871, to Helen Shelly, who died in 1878.

O'Connor had taken up the trade of making saddle trees, the wooden, rawhide-covered base for saddles, but by nature, he was an ardent cattleman, and he put his money into land and cattle. By the time the Civil War erupted, his herd was the largest in Refugio County (Grimes, 413).

When the Texas economy collapsed following the Civil War, O'Connor began selling his cattle, many through a packing plant in Rockport in which he may have owned an interest. These packeries proved to be one of the few commercially viable outlets for cattle at that time since markets for beef were far removed from the ranges. In this business, the workers slaughtered the cattle for their hides and fat, the latter providing tallow for making candles. There was not much of a market for beef preserved by salt drying, but trail drives took many cattle north to railheads in Kansas, and O'Connor and other ranchers also shipped cattle from the region's river and ocean ports.

As he turned his cattle into money, which was scarce in Texas at this time, O'Connor began buying land, an unusual and, to his neighbors, peculiar effort because most people ran their cattle on open range. But O'Connor's Irish background convinced him of the wisdom of owning property. By 1873 he had sold all of his cattle, and he used his profits to buy most of the land between the Mission and San Antonio Rivers. He then built a board fence between the two streams, and since the property backed on Copano Bay, this effectively enclosed the area. When the boards proved unable to withstand the fierce winds of gulf storms, he replaced them with barbed wire.

Some of those who had been ranging cattle on O'Connor's land before it was fenced had nowhere else to graze their stock, so they sold their herds to O'Connor. By the time of his death in 1887 he owned 100,000 cattle and half a million acres in Refugio, Goliad, Aransas, San Patricio, Victoria, McMullen, and La Salle Counties. These holdings qualified him as one of the Cattle Kings of Texas and made him "the wealthiest man in Refugio County" (Grimes, 413).

O'Connor's diverse interests included banking, and he was active in the firm of O'Connor and Sullivan in San Antonio. One of his major contributions was arranging to drill the first artesian well in the area. Although he died before its completion, he had the vision to seek the water essential to agri-

cultural efforts. He knew that with water and capital, a rancher has a chance to succeed.

After O'Connor's death, his two sons, Dennis M. and Thomas M. O'Connor, divided the properties. In 1861 Dennis had fought for the Confederate cause as part of the 21st Texas Cavalry, and he took part in the battles between the Red Legs (Union raiders distinguished by the red gaiters they wore) and Confederates in the Missouri-Kansas area. After the war, he ranched and joined the family banking business in San Antonio. He also contributed to efforts to make Rockport a suitable harbor. His generosity is apparent in his outfitting of the D. M. O'Connor Guards, Second Regiment, Texas Volunteers during the Spanish-American War. Dennis O'Connor died in Bryan, Texas, in 1900 in his private rail car en route to New York for medical treatment.

Dennis O'Connor's son, Thomas M. O'Connor, ranched and was active in business enterprises, including railroads and land development, especially at Port O'Connor. Perhaps his most significant accomplishment was his enormous contribution to U.S. cattle ranching. In 1916 O'Connor and Abel Borden, nephew and executor of the estate of legendary cattleman and trail driver Shanghai Pierce, imported fifty-one Brahman cattle from India. The herd was quarantined for an extended period in New York (Borden suspected rival ranchers were responsible for the delay), and some of the animals were slaughtered to be sure they did not have a communicable disease. Finally Borden secured President Theodore Roosevelt's help in getting thirty-three of the Brahmans released, and O'Connor received sixteen of them. This was the largest herd, up to that time, of Brahman cattle imported to Texas.

Brahman cattle, originally from India, where they are considered sacred, were first brought to Texas in the late 1870s. Five head—four bulls and a cow—were unloaded from a Dutch ship at the port of Indianola and purchased by Capt. John N. Keeran, a prominent rancher whose family still lives in the area, and Shanghai Pierce. James A. McFaddin, another prominent Victoria-area rancher, bought a bull in Louisiana about the same time. These ranchers believed that these disease-resistant cattle would be well suited to the Coastal Bend, a fact now well documented by time and practice.

One unusual characteristic of the Coastal Bend region was the number of black cowboys who worked there. After the Civil War, there were "as many black cowboys in the Coastal Bend as Anglos and Mexicans"; they learned the trade well and became "some of the finest cowhands that ever lived" (O'Connor, 6). Prominent black settlements included Lewis's Bend, the Black Jacks, and Sprigg's Bend. A strong oral tradition developed, and several black cowboys left written records that reflect a deep and abiding love for the ranching life and the families for whom they worked.

Today the properties owned by Thomas O'Connor's heirs include the Melon Ranch, Peach Mott Ranch, Duke Ranch, Salt Creek Ranch, and River Ranch. Combined, they encompass more than 200,000 acres. One note of interest is that part of the Salt Creek Ranch was purchased by the federal government and incorporated into the Aransas National Wildlife Refuge.

Ranching in the Coastal Bend region differs from that in other parts of the state in some interesting and challenging ways. Although forage for the livestock is usually plentiful, it is not as rich as grasses in other areas, such as North Texas, so animals require more than twice as much to stay in good condition. Cattle brought here from one of the areas with richer forage will lose weight, and even starve, because they are not accustomed to consuming the larger quantities of forage necessary to survive.

Parasites and disease are more of a concern here than in drier areas. Herds are constantly threatened by such ailments as red river disease, a usually fatal condition named for the red urine symptomatic of the disease; river lung disease; leptospirosis; anaplasmosis; foot rot; and other ailments, some of which can be prevented by vaccination.

Adverse climatic factors here include the high humidity, especially heavy dews most of the year, and salt air brought inland by a prevailing southeast wind common in coastal areas. The humidity and salt cause barbed wire, which in drier areas may last fifty years or more, to rust out in ten. Steel pipe used to construct pens rusts quickly unless carefully painted and repaired. Wooden posts rot, and metal fence posts rust away. None of the building materials have the permanence here that they do in drier climates.

The O'Connor Ranches are run by foremen Joe Keefe and Kai Buckert and about eight to ten full-time cowboys. The ranches also occasionally hire summer help. The crew rides almost daily and saddles mounts from a remuda of almost fifty quality ranch horses, mostly Quarter Horses of medium to large size. In the heat of summer, some of the men change mounts at noon. Others feel that it trains and conditions the horse to ride it all day. The only horses that have to be shod are those with hoof problems. The absence of rocks eliminates the need for shoeing, a chore that takes up a lot of time in rocky country. Although the ranch raises some horses, they produce no horses for the race track, show ring, or cutting arena.

The saddles, which are furnished by the ranch, come from various makers and have rolled cantles and large, high front swells, often extending beyond the rider's legs for protection from the brush. Only occasionally do riders use tapaderos, except on the land in Goliad County, which borders the southern edge of the O'Connor property and has some typical South Texas brush. There tapaderos are necessary to protect the rider's feet from thorns. The bits are the usual shallow port type; riders use both split reins, found in much of North and West Texas, and the single looped rein common in South Texas. Some reins and other pieces of tack are made from woven or plaited nylon rather than leather because the synthetic fibers outlast the leather in the dampness.

One noticeable difference on the O'Connor ranches is the length of lariats, which is usually forty feet instead of the thirty to thirty-two feet found in most of Texas. Kai Buckert, one of the foremen, likes the longer ropes because the Brahman-cross cows "will come up the rope to meet you," something that makes a roper want a little more space to work with. Ropers usually tie the lariat to the saddle horn hard-and-fast, although some dally the rope around the horn. On the brushy southern properties, a rider has to be particularly careful about tying onto a large animal and entangling the horse and cow in the brush.

The ranch of Louise O'Connor is a cow-calf operation with about 2,500 crossbred cows and a few hundred heifers kept for the breeding herd each year. Some of the animals have the blood of Santa Gertrudis, that breed developed by King Ranch to withstand the conditions found on this range. But most of the cattle are a cross between purebred—but not registered—Hereford cows and Brahman bulls. The purebred Herefords are kept separate from the crossbred cattle raised for sale, and mature Brahman bulls are bought from outside sources. By following its breeding practices, which rarely have included artificial insemination, the ranch assures a first-line cross between its two primary breeds.

The cattle carry the brand TC, which belongs to Louise O'Connor and Kathryn O'Connor Counts for their partnership, O'Connor Brothers Ranches. The ranch operates the brand somewhat differently from other ranches. After the heifers that are to be kept for breeding have been selected, they are branded. The other calves are sold without brands. No cattle bearing the ranch brand are sold, except for slaughter. The ranch also uses other brands, including the "a" little "a" brand on the Melon Creek Ranch, the lazy J on the Williams Ranch, and the buggy pole on the ranch of Maude Williams. The triangle dot brand is used on the Salt Creek Ranch.

Alejandro De La Garza

The annual cycle of work begins in January when the men castrate, vaccinate, and earmark the calves by sex, the latter step assuring that heifer and steer calves can be identified easily later. The men separate the calves from the cows in a cutting alley and push the calves into a small pen where the men wrestle the animals down. In February the bulls are put out with the cows, where they will stay until August, a longer period than the sixty to ninety days found on some ranches. In March the yearling heifers are vaccinated to prevent disease, and the cow herd is vaccinated by mid-April. The animals are also sprayed for flies during these late March and April roundups. In May and June there is another round of marking calves and spraying for flies. By June and July the offspring from the first-calf heifers go to market. By September the big calves go to market, and the cowboys cut out and wean the heifers to be kept. As the fall progresses, the other calves are gathered, sorted, and shipped to a buyer who had visited the ranch earlier to place a bid. The successful bidder sends trucks for hauling at a time agreed upon by the buyer and ranch personnel.

The ranch hands are of Anglo, Hispanic, and African American descent. Some Spanish is used to give directions, but the principal language is English. The cowboys live in houses near the headquarters and haul their horses in gooseneck trailers to wherever they're working that day. This eliminates the need for anyone to live in remote areas on the ranch.

One of the major chores, other than working cattle, is seeing after the stock's water supply. The ranches have more than a hundred wells, some of which are pumped by the sixty-five windmills or electric pumps. However, most of the water flows from artesian wells that in this area are about nine hundred feet deep. Some of the water is contained in earthen tanks, but concrete troughs are more common. The time and expense required for pumping water and repairing windmills are major concerns for the ranch.

Pens located on the ranch are often constructed of wood; some are of pipe and steel cable. Constant attention must be given to repair and replace fences to ensure that they will hold the cattle when the time comes for working them. The horse barn on the River Ranch has walls of cast cement with pipe gates, even though the corral is made of two-by-six planks with plank gates.

Winter feeding is less of a chore here than in colder climates, but the ranch hands do feed hay. The pasture grass retains some food value, but the ranch must supply some form of protein supplement. The hay cut and baled on the ranch is an economical way to do this. Square bales are given to the horses and cattle that are penned, and large round bales are fed to pastured cattle. Some limited planting of oats for winter feed is done.

Some of the finest grasses for cattle are found in the area—bluestem, grama, switch, Dallas, and others. The southern ranches have smut grass, Bahía, common Bermuda, and some coastal Bermuda. Brush includes mesquite, running oak, persimmon, huajillo, and catclaw, all characteristic of the Brasada of South Texas. Controlled burns, bulldozers, and sprays are used to control these plants.

Wild game is plentiful and includes deer, javelina, bobcats, and quail, but the ranch does not lease land for hunting. Also found here are long-tailed cougars or mountain lions.

Several significant structures stand on this ranch. One of these is the original house of the O'Connor family located where the log cabin was erected in 1842, now on the River Ranch. The house was built in 1874 of pine lumber from Louisiana brought by ship to the opening of Matagorda Bay. Unfortunately, the ship arrived at low tide, and the captain had the crew throw the lumber into the surf so that it would be carried to shore by the tide. O'Connor nearly rejected the shipment but decided to accept it when he saw the quality of the wood. It was almost all heart of pine and is still sound these many years later.

When one of the boards requires repair, holes must be drilled because the wood is still quite hard.

Another structure is a Spanish-style Catholic church built by Kathryn Stoner O'Connor in 1952. Saint Anthony of the Woods, the original church on the ranch, was destroyed by a storm in 1942. This storm devastated structures on the ranch, but because of World War II, materials for building, even repairing, were not available. When Mrs. O'Connor's son Dennis entered military service, she made a promise to God that if he returned safely, she would build another church. Built in the style of a Spanish mission, the church's thick limestone walls are a barrier to wind and sun. To avoid flood damage, it was built farther away from the river than was Saint Anthony of the Woods. The structure's unique lower walls were built with large blocks cut from heavily fossilized stone, giving the walls a lot of texture. The church, with its traditional long main hall and side rooms, has the form of a cross which can be seen clearly from the air. Family members, ranch hands, and people from the surrounding area attend Mass at the church.

Other interesting structures include a kiln and a camp house. The kiln was used to make all of the brick used on the ranch, much of it for sidewalks. The kiln is more than a hundred years old, dating back to a time when bricks were not available for purchase. The old camp house is a historical text of sorts because cowboys over the years have carved their initials and dates into its wood.

The O'Connor family continues to be dedicated to ranching. Dennis M. O'Connor II was only two years old when his grandfather tied him to a saddle. He recalls, "As soon as I was big enough, I went with the ranch crowd. That was play for me. I always wanted to ranch; I was raised here. Even though I was prepared for a career in chemistry, it never occurred to me to be anything but a rancher. I do it because I love to work cattle, not because it is my responsibility" (O'Connor, 55).

Ray Ranch

THE IBEX LAND AND CATTLE COMPANY RANCH AND THE NOLAN RYAN RANCH

Two ranches currently operated as the Ibex Land and Cattle Company Ranch and the Nolan Ryan Ranch originated as the Ray Ranch. This spread is located about halfway between San Antonio and Corpus Christi, and just south of Choke Canyon Reservoir. Tilden, the seat of McMullen County and currently its only town, lies to the west on Texas Highway 72, which forms the north boundary of the Ibex Ranch. These ranches are in the eastern part of the county along the Live Oak County line, and in the northern part of the South Texas Brasada or brush country. Although almost a hundred miles from the Gulf Coast, the area's elevation is only about 150 feet above sea level.

The seemingly impenetrable barrier of thorny brush makes excellent cattle country despite the dearth of grass. Cattle browse on the various plants, especially the huajillo, and in summer they fatten on the protein-rich beans of mesquite trees. Frost during the winter is rare this far south, so plants grow almost year-round. This is a hostile, demanding, but still productive region where drought is common, as is periodic heavy rain from storms coming ashore from the Gulf of Mexico.

In his *Tales of Old-Time Texas*, J. Frank Dobie tells the story of the founding of this ranch. The saga began in Alabama following the Civil War with Elijah Ray and his business partner, a man named Hess, as well as Ray's two sons, Wallace and Jim. After he and Hess sold a store in which they were partners, Ray gave each son a custom-made money belt containing five thousand dollars in gold. When the pair decided to go west to establish a ranch in Texas, Ray made sure that each son carried a six-shooter, a rifle, and extra clothing. Their mother made sure that each had a Bible.

The young men traveled down the Mississippi River to New Orleans and from there to Indianola, the port that served as the point of entry for many early-day settlers who came to the Texas coast. The town was located near present-day Port Lavaca, on Matagorda Bay. (Both the port and town of Indianola were destroyed by storms in the 1880s.) The brothers outfitted themselves with horses, Texas saddles, leather chaps to protect themselves from the brush, and the traditional boots and spurs. They traded their long rifles for saddle carbines, bought extra cartridges and food, and headed out.

Although they heard in an Indianola wagon yard that their father's former business partner had passed through the area, they attached no significance to the news. The two young men traveled north to Beeville and then on to the Nueces River country near the current towns of Three Rivers and George West. The two were impressed with the countryside, and also the cattle, many of which they noticed were not branded. Their situation seemed promising. Their apparent good luck, however, turned bad at this point.

While riding through the ranch country, they were attacked by a band of riders, all of them Indians except one. That one proved to be Hess, their father's greedy former business partner, who was determined to get the gold that he knew each of the boys was carrying. How Hess had fallen in with the Indians no one ever knew. In the chase that followed, Wallace was killed, but Jim abandoned his horse and hid in the brush. By this time he had only one unfired cartridge in his six-shooter. His rifle was still on his horse.

Early the next day he found the Indian encampment and recognized Hess. After standing in the open to be sure that the Indians saw him, he quickly darted into the brush and managed to elude them. He then moved back upstream, recaptured his horse and supplies, and rode away before his pursuers returned for their extra horses and gear.

After he was sure the raiders had left the area, Jim went back to bury his brother. Realizing the danger he faced in open country, he hobbled his horse near grass and water, and traveled down the stream on a crudely constructed raft. Along the way he discovered the remains of Hess, who had drowned while fording the swollen stream. Jim removed Wallace's money belt from Hess' body, which had probably weighted him down, making him unable to swim in the swift water. Then the weary, disconsolate man continued his journey. The raft soon broke up, and Jim crawled up the bank. He hailed a man on horseback, who gave him a ride to his nearby ranch.

As Jim recovered from his ordeal at the ranch of his rescuer, he discovered that the land he had noticed on his odyssey was state land subject to sale or homesteading. Using his and Wallace's money, Jim bought horses, a wagon, and other necessary provisions; hired help to build a house; and established the Ray Ranch.

The ranch has been run by members of the Reagan family since June 1919, when it was leased by the family of Rocky Reagan, a South Texas cattleman of considerable note. That year he moved to the Ray Ranch from the Shiner Ranch near Tilden with his wife, Eula Cleveland Reagan, and their children. A noted storyteller in his own right, Reagan related to J. Frank Dobie the account above of the founding of the ranch. Reagan eventually published several volumes of tales and a biography of his country doctor father.

Well before Rocky Reagan's death in 1975, his son Bob took over the lease and ran the ranch until the property was sold to Edwin Singer and Fred Roach

Ray Ranch foreman Bob Reagan

and Company in 1967. Bob became manager of the part of the ranch purchased by Singer, which in 1997 was bought by the Ibex Company. Bob still runs it today, despite being more than eighty years old. He is assisted by his granddaughter, Nicki Dye Martinez, who rides in the brush like a vaquero.

Bob Reagan recalls his early years on the ranch as a colorful period filled with the spirit of range life. In 1919 the ranch was fenced only on the boundary lines. It was, in effect, a thirty-thousand-acre pasture choked with brush. Although it has since been crossed-fenced, and some efforts have been made at land improvement, it remains today, as Bob says, "a brush country ranch." Anyone accustomed to ranging cattle on tall grasses in open country would be lost here.

Working the cattle in the early part of the century were men who called the chuck wagon home for much of the year. Bob recalls that the ranch's remuda of saddle horses at that time numbered 150 head who were looked after by a "remuda man." Each vaquero had a string of twelve horses, and when the wagon went out, each one took six of those horses with him. Each day the men changed mounts at noon. In about a month, when their horses were exhausted, the crew would return to headquarters to exchange their tired horses for fresh ones. In order to supply the remuda, the ranch kept a band of forty mares.

The ranch no longer raises horses, preferring to buy them instead. Bob Reagan is responsible for these purchases and looks for medium-frame Quarter Horses because they are sturdier than those with more slender builds and are less easily injured in the demanding work. He avoids large horses in preference for horses 14.3 to 15.1 hands high. These horses are large enough to carry riders and withstand heavy work such as roping and holding wild cattle, but nimble enough to work in the close quarters imposed by the brush. Riders use both geldings and mares, not a universally accepted practice in ranch country, but in a day when few breeders raise horses, it does make it easier for Bob to find mounts for his vaqueros.

To facilitate the herding of cattle prior to shipping, Rocky Reagan fenced a two-thousand-acre trap in the middle of the ranch soon after his arrival. The Nueces River runs through the trap, so the cattle had ample water. At shipping time, the hands gathered the stock from the various parts of the range, cut out the cattle to be shipped, and drove the selected animals to the trap to be held until the gathering was completed. Having a secure trap eliminated the need for night herding. Once the herd was complete, the men drove the cattle out to the road through Simmons City (now a ghost town) and on to the railroad corrals in Three Rivers, where the cattle were loaded onto rail cars and shipped to Kansas City.

Many of the people who lived in the area during this time have vivid memories. Elmer House, one of the old-timers in a family long associated with the area, recalls many cattle drives to Three Rivers in the years leading up to World War II, before trucking cattle to market became the preferred method. Dauris Mahoney recalled how, when loading the Longhorn steers in the rail cars, the men had to turn the cattle's heads in order to fit their horns through the doors. A young clerk at the Matkin General Store in Simmons City and son of a local rancher, remembers driving imported Mexican steers from the railroad to the ranch, often a wild ride for the hands. After they were fattened for market, the steers were driven back to Three Rivers to be shipped.

Working calves in this part of the state is customarily done without using chutes or roping and dragging. After cutting the calves away from the cows, who might attempt to gore anyone hurting their calves, the vaqueros simply hem the young animals in a small pen, throw them to the ground, and work them. This appears to be a regional tradition in South Texas and likely grows out of the Mexican vaquero tradition. In West Texas riders either head or heel the calves and drag them to the branding fire, or, on some ranches, use working tables.

Local fauna include whitetail deer, feral hogs, and javelina as the main game animals, with doves and both bobwhite and blue quail being popular game birds that thrive there. Exotic animals pastured on the Ibex Ranch include elk and black buck antelope. Typical predators are coyotes and bobcats, though mountain lions occasionally prowl the region. Hunting rights are not leased to the public. A common hunting method in this type of country is to cut senderas through the brush and then sit in stands that overlook these open trails.

Most of the fences have five strands of barbed wire, but some ten-foot-high net wire fencing is used, particularly along the airplane landing strip near the Ibex headquarters. The pasture holding the exotic game also has high net wire fencing. Cedar posts have given way to metal posts for new construction.

Riding in the brush requires some adaptation, and some riders use short leather tapaderos on their stirrups to protect their feet from the thorns, but not on the Ibex Ranch. The ranch furnishes the vaqueros with double-rigged Texas-style saddles, not the traditional Mexican saddles. These have swelled forks and rolled cantles. The men do, however, protect their legs with shotgun chaps, which Dobie insists are referred to as "leggins" in this part of the state (*Vaquero*, 193). These are fashioned from heavier leather than that used for chaps worn in less-threatening terrain. "Chinks," or short chaps, do not offer enough protection to gain favor here.

The ropes are usually thirty feet or less. The heavier the brush, the shorter the rope. The earlier use of grass or manilla ropes has given way to stronger nylon ropes. A heavy canvas brush jacket and leather gloves are necessities for this kind of riding. Since the Ibex Ranch furnishes the gear and tack, the choices reflect the region, not the vaquero influence that might be apparent if riders furnished their own gear.

The vaqueros today, as they have been for years, are native Mexicans who are legal immigrants. Hence the working language of the ranch is Spanish. The vaqueros work from mid-January to mid-December, when they go back to visit their families for a month before starting the annual cycle of work again. They live in a bunkhouse and eat meals prepared by the ranch cook. The Ibex Ranch does not use helicopters, so even though the pastures are large (the largest is two thousand acres), the cattle are gathered by vaqueros on horseback. The vaqueros are sometimes referred to as "brush poppers" because of the sounds they make as they ride full speed through the brush, breaking limbs in their wild pursuit of the cattle. Old-timers say that when a good rider breaks out of the brush after chasing a wild cow, he will have enough wood caught around his saddle horn to make the fire to cook his supper.

The cattle are crossbred with a Beefmaster base, and most of them show evidence of Brahman blood. Since the Brahman, originally from India, is genetically suited for the hot, humid climate, the breed has become almost universally accepted along the Gulf Coast. The cattle have no Longhorn blood.

Forage for the cattle is unusual to those unfamiliar with the region. Prickly pear and huajillo are the two principal food sources for stock. In order for the cattle to safely eat the prickly pear, ranch hands burn the spines off with a propane torch. They often use a 250-gallon propane tank mounted on wheels and equipped with one or more fifty-foot-long hoses. Cattle apparently enjoy eating the fleshy pads still warm from the burner. Lighting the roaring torches is certain to lure the cattle out of the brush and start a feeding frenzy like those triggered in other regions by the distribution of range cubes, hay, or other sorts of feed, especially in winter. Rocky Reagan, his son Bob recalls, thought that with each "norther," or

winter storm, the crew needed to get out the pear burners and set to work.

Since the prickly pear here grows in large clumps that may stand four or five feet tall, it seems more practical to feed cattle the singed prickly pear than it does in other regions where the pear grows close to the ground with dry grass stands at the base. In those regions, it is far too easy to start range fires, and the practice of burning pear has never become popular.

Because the cattle depend on the brush as a food source, control of it is limited to some removal of mesquite and other parasitic growth. Bob Reagan said, "I don't want all of the brush removed; it makes up most of our grazing." In some areas that have been cleared in the past, it is necessary to cut back the brush periodically with a shredder pulled by a tractor equipped with thirty-ply tires to prevent puncturing from thorns. Controlled burning of brush has not caught on in this area as it has in other parts of the state. The practice is ineffective in areas where there are insufficient weeds and grass to sustain a hot fire that controls the brush and weeds and enhances growth of grasses.

Efforts to improve forage continue. The Ibex Ranch is experimenting with leucaena, a plant from Hawaii that seems to adapt to South Texas climate and resembles huajillo. When harvested, it is processed into pellets and used for feed, especially for deer. It has a protein content of 27 percent, unusually high for a plant, making it one of the most promising developments in the area, but it is raised only on irrigated fields for feed at present.

Water on the ranch comes from collecting runoff flow in surface tanks. Since no subsurface water is available, no windmills dot the landscape. Some subsurface water exists farther south, but it has a heavy mineral content and is fit only for livestock. The Nueces River flows for about five miles through the ranch, and in 1971 Bob Reagan built a dam that backs water up about ten miles and provides a reliable supply of water for cattle and game. Although the ranch has an irrigation permit, the water is not used to farm. Farming in this dry region is risky, although growing grains and other crops is common. The caliche soil becomes hard when dry, although the black soil along the river is rich and fertile. If the land is plowed, the ranch hands must plant desirable crops, or weeds take over and render it useless to cattle and game.

The owner of the north portion of the ranch is the Ibex Land and Cattle Company of Pittsburg, Texas. The other part of the Ray Ranch is owned by baseball great Nolan Ryan, who brands an N-R. Long active in agriculture circles, Ryan has several properties in the state and recently purchased a motel and restaurant overlooking Choke Canyon Reservoir.

The ranch stands today as a model of what could happen to other large ranches. Over time, it became necessary to sell the ranch, which was too large to be bought by an individual; therefore, the ranch was divided in half and bought by people whose money came from outside ranching circles. This seems to be the future of other large ranches, which will likely be broken up as succeeding generations pass through the continuum of time.

Panhandle and Northwest Texas

NORTHWEST TEXAS AND THE PANHANDLE ARE HOME TO several of the most famous ranches in the state. The ranches included here lie for the most part west of the Brazos River, south of the Red, east of the Pecos, and are cut by the headwaters of the Colorado. Two, the JA and the Goodnight, are found in the Texas Panhandle, most of which is an extension of the Great Plains. One of the most notable geographical features of the Panhandle is the Canadian River, along whose wandering course ranching has flourished. The most distinctive feature, however, is Palo Duro Canyon.

In the 1800s, this was buffalo country and home to the Comanche and Kiowa Indians. The rolling hills and plains extend to Cap Rock, where the High Plains of the Panhandle begin. The XIT Ranch once covered three million acres here before it was sold and split up into smaller ranches that are still large, even by Texas standards. The famous JA Ranch, the oldest privately owned ranch in the Panhandle, is a flagship of Texas ranching.

Cutting through the otherwise vast level plain of the Llano Estacado, or the Staked Plains, Palo Duro Canyon stretches for sixty miles along the course of the Prairie Dog Town Fork of the Red River. It is six miles wide in places, and the sheer walls rise eight hundred feet or more from floor to rim. The Spanish name of the plains stems from the stakes driven by Spanish explorers used to navigate through the otherwise featureless landscape. Fossils in the canyon walls reveal four different geologic periods and range over 240 million years. Human activity in the canyon dates from as early as 10,000 B.C. Once the home of herds of buffalo and other wild game that sought the well-watered grasses and the protection of the steep walls from icy winter winds, it was also the winter refuge of Plains Indians until forces led by Col. Ranald Mackenzie surprised them there in 1874 and burned their winter supplies. The soldiers also captured the horse herd and purposely destroyed more than a thousand of them, thus ending forever the reign of a horse culture that produced some of the best light cavalry the world has ever known. Palo Duro Canyon is the state's largest park, including 16,000 acres in

Armstrong and Randall Counties. The park is the home of the annual production of *Texas,* an outdoor drama depicting settlement of the region. The area is still heavily devoted to both farming and ranching, and oil production is found in much of the region. Large areas of the Panhandle have been made productive by water pumped from the Ogallala Formation that underlies much of the plains.

The absence of timber challenged early settlers accustomed to log cabins and led to the construction of alternative structures. Common in areas with a limited supply of typically small trees was the picket house, erected by driving the sharpened ends of the small logs upright into the soil to form walls and chinking the cracks with mud. In areas with no timber, settlers used dugouts, actually little more than a hole dug into the side of a hill with a front wall and roof.

Grasses in the area are rich and plentiful, except in time of drought. These include Johnson, buffalo, Bermuda, mesquite, grama, sprangle-top, rye, and fescue. These grasses are very nutritious, and with the low humidity, they dry well in the fall and remain nutritious into the winter.

Not nearly so dense as the Brasada, the area south of the High Plains is nonetheless plagued with brush, especially mesquite, prickly pear, tasajillo, and other thorny growth. Efforts to control it include removal by machines, controlled burns, and chemicals.

Game is plentiful in Northwest Texas and in the Plains canyons, and includes whitetail deer, bobwhite quail, turkeys, doves, and feral hogs (but no javelina). Bobcats and coyotes are plentiful, but cougars are extremely rare.

The Western Cattle Trail, which cut through the Rolling Plains, was popular with drivers after the Chisholm Trail, which ran near Fort Worth, lost popularity. Because of settlement in Kansas, cattle drives were shifted further west to avoid damaging Kansas farmers' fields and transmitting tick fever to their cattle.

This area remains sparsely populated except for the major centers of Dallas, Fort Worth, Abilene, Wichita Falls, Lubbock, and Amarillo.

R. A. Brown Ranch

THE HEADQUARTERS OF THE R. A. BROWN RANCH LIES WEST of the town of Throckmorton, a small ranching and farming community in Throckmorton County between Abilene and Wichita Falls. The current owner and operator of the ranch is Rob Brown, the third generation of the family to run the ranch, and his wife, Peggy Donnell Brown. Along with their children and the children's spouses, they currently operate this very productive cattle and horse operation in some of the best ranch country Texas has to offer.

The Brown family has been ranching in Texas for more than a century. They trace their ancestry back to Joseph E. Herndon, who came to Texas in 1857 and settled eventually in Robertson County. He operated a plantation that also included some livestock. A son, J. W., was killed at Gettysburg during the Civil War, but his daughter, Lucy, married Robert Alexander Brown, a merchant in nearby Calvert who had a business in Galveston after his return from the Civil War. Brown took over the Herndon plantation and in 1876 bought the historic jail in downtown Calvert and turned it into a dwelling. Family records indicate that R. A. died in 1890. His and Lucy's oldest son, Robert Herndon Brown, married Belle Scott in 1889 and began ranching in the Waco area soon thereafter.

R. H. Brown moved his family to Jack County and founded his Bar O Bar Ranch, which he sold in 1903 to pursue a different kind of career, one nonetheless associated with the livestock business. Brown moved his family to an elegant Victorian-style house on Ballinger Street in Fort Worth and managed Evans, Snider, and Buel Commission Company at the newly opened Fort Worth Livestock Exchange. Still drawn to ranching, however, he started buying land in Throckmorton County, west of his former holdings in Jack County. The land he bought has rolling hills, with draws and occasional creeks. Limestone is common and serves to make the forage rich for grazing livestock. Early settlers recall that the area was covered in grass "stirrup-deep on a horse" with trees only along the watercourses. Brown's partner was W. A. Poage, a rancher who lived near the town of Woodson in southwestern Throckmorton County. Together they amassed sizeable holdings, which they later divided. Poage kept the original headquarters near Woodson, and the Browns established a new headquarters just west of Highway 283, west of Woodson.

The man who took over the ranch and developed it into one of the best livestock operations in the state was R. A. Brown, the son of R. H. and namesake of his grandfather. R. A. was born on December 7, 1902, on the Bar O Bar Ranch and moved with the family to Fort Worth. He began spending time at the ranch as a child, especially after his father became ill with cancer in 1919. When he was fifteen, he moved to the Throckmorton County ranch to make it his home.

Times were difficult economically because of financial depression and drought. To help offset these factors, R. A. helped skin cattle that had starved on the parched ranges. He then loaded the hides into a wagon and hauled them to Abilene, some sixty miles distant, to sell them and buy cottonseed cake, a protein supplement that was the standard feed at that time. Such hides were often referred to as "poverty hides" because of the hard times that gave rise to this activity. This widespread loss of livestock was eventually avoided when ranchers had the resources and opportunity to store feed for the winter and periods of drought.

Brown was always drawn to horses and began playing polo as a young man. In 1922 he entered the Agricultural and Mechanical College of Texas, now Texas A&M University. During his stint of a year and a half, he played polo in a collegiate setting. He was forced to drop out of school because of his father's failing health and economic hardship. This ended his higher education but did not diminish his love for good horses and polo.

In the late 1930s, R. H. died, and R. A., his mother, and his sisters inherited the ranch. His mother and sisters turned control of the property over to R. A. in hopes that he could save it from being repossessed for debt. Although warned that there was little hope of saving the operation from foreclosure, R. A. consolidated the properties, borrowed more money, and began operating the ranch.

Not all of his activities were rewarding. One of his early ventures was the purchase of a small herd of steers one fall. The animals wintered well, but just as they appeared ready to market at what promised to be a profit the next spring, disaster struck. During a thunderstorm, several of the animals bunched up in a corner, and when lightning struck the fence wires, a number of them were killed. R. A. saw his profit disappear. Despite setbacks of this sort, by 1941 R. A. had paid off the debt, and the economy had begun to improve as the country geared up for World War II. R. A. then redivided the properties among his sisters.

One activity through which R. A. obtained operating capital and cattle was helping an uncle in the business of foreclosing on ranchers who had to default on loans because of the Great Depression. On one particular trip in 1933, he traveled with a small herd of horses to a ranch on the New Mexico line. He took possession of a herd of mortgaged cattle on a ranch suffering foreclosure and hired a crew of cowboys to drive them back to Throckmorton. Decades later, the family that had defaulted on its loan and lost its herd, but not its ranch land, traveled to the

Browns' annual sale to buy bulls to run with their new herd.

On November 14, 1931, R. A. married Valda Thomas, daughter of Mr. and Mrs. D. B. Thomas of Throckmorton. The young couple had two children, Marianne and Rob. Marianne was born in a small house still standing on property controlled by the Brown Ranch, between Albany and Throckmorton. Although Valda was happy with her marriage, the life was difficult and demanding. The house had running water but few other modern conveniences for a young mother with a baby to tend. Valda remembers relying on kerosene for both cooking and lighting. There was a fireplace for warmth, and also a gasoline heater, but she was afraid to light because it could flare out of control when first lit. She counted on ranch hands to make that source of heat available when they could.

R. A. soon established the ranch as a breeder of quality Hereford cattle and fine Quarter Horses. He was instrumental in bringing some of the earliest Hereford cattle into the area and carefully breeding these animals to supply both beef and breeding stock. Although later he would become interested in crossbreeding, throughout his life he strongly believed in the virtues of the Hereford breed.

R. A. also had a keen interest in Quarter Horses, which are smaller and quicker over short distances than Thoroughbreds. He had developed this appreciation playing polo, which he was able to do again after he recruited some cowboys from nearby Woodson to form a semiprofessional team. Since they had no trailers in those days, they would drive their horses to Abilene or Wichita Falls (more than sixty miles) one day, play the next day, and drive them back on the third day. It took dedicated players to hold up to that regimen.

During most of his years of running the ranch, R. A. kept about eighty broodmares. Some of these were descended from cavalry horses from the nearby remount station, but R. A. introduced other excellent bloodlines into his herd. He traded for Black Hancock, offspring of the famous Joe Hancock, from the 6666 Ranch, where he later purchased Blue Gold, a gray horse by Blue Rock out of a famous Hollywood Gold mare. He also bought Joe Bailey Rickles, a double offspring of Weatherford Joe Bailey.

The ranch produced many fine horses. In 1946, R. A. got what was for the time an exceptionally high price for the two-year-old filly Brown's Firefly, the grand champion halter mare at the Denver Stock Show. The Haythorns of Nebraska bought her for five thousand dollars. Haythorn had a young stallion that had won his class, and he wanted to breed these two exceptionally fine animals. A close relationship continues between the R. A. Brown and Haythorn Ranches.

R. A. helped found the American Quarter Horse Association in 1941 and later served as vice president. After his death, an article in the *Oklahoma Quarter Horse Breeders Association* magazine stated that few, if any, had "tried to do more for the betterment of the American Quarter Horse Association than the late R. A. Brown" ("R. A. Brown," 16). The breed's popularity in the United States as well as in South America and in Europe proves the wisdom of the founders. The association has registered more than three million horses in its history.

R. A. was also interested in greyhounds, coon dogs, and bird dogs, and he became a prominent breeder of quality hunting dogs. He enjoyed using the greyhounds to chase coyotes and even had a truck specially rigged with cages with quick-release doors to enable his dogs to get the jump on coyotes spotted as he was driving in his pastures. Valda recalls having as many as thirty dogs in pens behind their house. Often R. A. would go off to sell some of the dogs, but he usually returned with others to replace them. Valda hunted with her husband, many times on horseback, and because she usually rode the gentlest horse, she often had the responsibility of carrying the raccoons, opossums, and other small animals that the dogs hunted down. The Browns skinned these and sold the hides.

R. A. also became a leading proponent of sound land management. In the 1950s he began grubbing mesquite trees and clearing out prickly pear, the area's major parasitic plants. This practice would become more popular, but was uncommon at the time. In one pasture R. A. left only one large mesquite tree near the road as a conversation piece.

When R. A. died in 1965, management of the ranch passed to his son, Rob. Today, the operation includes the Headquarters Ranch west of Throckmorton, the leased Crooked River Ranch, another leased ranch near Graford, a ranch near Eliasville inherited by Peggy Donnell Brown, and a ranch in Colorado near Matheson. Rob, a 1958 graduate of Texas Tech University with a degree in animal science, and his wife, Peggy, herself the daughter of pioneer stock in the region, moved the ranch into the modern era by instituting a number of changes in operation.

Strongly influencing those changes was Dr. Dub Waldrip, a former manager of the Texas Experimental Ranch, an operation of Texas A&M University between Throckmorton and Seymour, and also for many years chief operating officer of Spade Ranches. Because of his influence, Rob began actively crossbreeding cattle, continued breeding horses, became active in range management, and moved into public sale of Brown Ranch livestock.

As early as 1965 Rob had begun experimenting with crossbreeding Herefords. First he worked with Brown Swiss cattle in order to develop a cow that would produce more milk to grow a better calf. When he found that buyers were not willing to pay top prices for the brindle and sometimes strangely colored offspring, he looked further and discovered the merits of the Simmental, a breed whose red coloration fits more closely with the patterns of the Hereford. He began extensive crossbreeding with Simmental and in 1974 became president of the American Simmental Association. During his tenure, the breeders group added three thousand new members, and Simmental now ranks as the third most prevalent registered cattle breed in the United States. The Browns also breed Red and Black Angus, as well as Simbrah and Senepol, and have become highly respected suppliers of excellent bulls. The ranch has risen to ninth in the United States as suppliers of breeding stock of cattle and ranks as ninth in the nation in registering Angus cattle.

Rob continued the active breeding program that by the late 1960s ranked among the top twenty-five producers of fine Quarter Horses in the United States. Although the ranch did not compete in cutting, halter competition, or racing, the horses they sold to those who did compete in these activities established the ranch's reputation. Heeding his mother's warning about how much time R. A. had to spend with his horses, Rob reduced the number of broodmares to around thirty. He continues to pasture-breed the horses, a good way to raise ranch stock, and sales of his horses continue to be strong. The ranch regularly sells horses to breeders in Hawaii, Europe, and South America. The ranch has also won the Best Remuda Award from the American Quarter Horse Association and National Cattleman's Beef Association. This award, given for the first time in 1992, honors the top breeders of ranch horses in the United States.

Rob travels widely in support of the work of the American Quarter Horse Association, frequently in South America and Europe. He served the association as an officer for several years and as president in 1995. In Germany, especially, Rob has found strong interest in the Quarter Horse, which is much different than the larger horses typically found there. One German rider compared the two types of horses by likening the larger European horse to a truck and the Quarter Horse to a Mercedes sports car.

A good cow horse must be not only well bred but also well trained. Rob has worked out an arrange-

EST 1895
RA BROWN
RANCH

ment with his cowboys to help in this lengthy process. If a man does a good job training a horse and agrees to part with it when the animal is in its prime at seven or eight years of age, the cowboy is given 25 percent of the income from its sale.

Rob became extremely active in range management, and in 1974 he received the Outstanding Range Management Award from the Texas section of the American Society of Range Management. He has carefully used sound grazing practices such as pasture rotation, controlled burns, and brush removal to increase the grazing capacity of the land.

In order to market their livestock more effectively, Rob began holding annual sales of bulls and horses each fall in a social and business event that has grown in popularity over the years. In early October the stage is set as the cowboys gather and pasture the animals close to the headquarters west of Throckmorton. The men sort and pen the bulls for examination by potential buyers on sale day. Buyers can also review bloodlines and performance data in a meticulously documented sale catalog.

One of the sale's unusual highlights is the display of weaned colts and fillies for sale. These are haltered and tied to overhead cables in a large pen where potential buyers can move freely among the wary young animals. At six months of age, these delicate, intelligent creatures show great promise for careers as breeding stock or for working cattle in the cutting arena or pasture.

Ranch operations involve more than the activities and efforts of the owners. It takes good help to run a successful ranch, and that means good cowboys mounted on good horses. In this the Brown Ranch has been blessed, because they have worked closely for many years with members of the Self family. George Truett Self and his brothers Carl and Pete began working for R. A. Brown more than sixty years ago. Truett's son George also works for the ranch.

Rob and Peggy's children and their spouses also assist in running the operation: Betsy and her husband, Jody Bellah; Marianne and her husband, Todd McCartney; Rob A. and his wife, Talley; and Donnell and his wife, Kelli. Both Donnell and Kelli were national presidents of Future Farmers of America, and Donnell and Jody each have won the Top Hand Award at the Texas Ranch Roundup, the major ranch rodeo held each August in Wichita Falls. This contest pits several major ranches in the state—Waggoner, 6666, Pitchfork, and others—against each other, and the Brown Ranch has a long history of successful competition.

The Brown Ranch continues to attract attention as a quality family-run operation that has diversified and continues to improve its operation using mainly family management and excellent cowboys. In 1994 the National Cattlemen's Association honored the family as part of the celebration of five hundred years of cattle in the New World. It is the continued existence of operations such as the Brown Ranch that proves family ranch life in Texas is alive and well. In 1998 the ranch was recognized by the National Cattlemen's Association as Cattle Business of the Century.

Chimney Creek Ranch

CHIMNEY CREEK RANCH LIES NORTHEAST OF ABILENE, FIVE miles west of the highest point in the region, at which Highways 6 and 351 join. To the northwest is Stamford, home of the famous Swenson Ranch. The land here was once the range of the buffalo and the Comanche Indian who depended on the shaggy bison for food and shelter. Since the demise of the buffalo, it has been cattle range cut into numerous ranches both large and small. Chimney Creek Ranch is mostly rolling plain, but to the east the land is broken by canyons and draws. The rocky region is not level enough for farming. Nearby Albany, which calls itself the home of Hereford cattle in Texas, has been a center for the ranching culture that settled in this area in the late 1850s. Early settlement was sparse because of the threat of raiding Indians and a scarcity of water. The major source of water for livestock and people was, and still is, tanks hollowed out in the draws to catch runoff from the infrequent rains. Some wells along creeks and draws provide adequate amounts of water for livestock, but the windmill which helped settle the High Plains is absent from this landscape.

Settlement in the area before the Civil War was aided by the establishment of Camp Cooper almost thirty miles to the north on the Clear Fork of the Brazos River. Here Col. Robert E. Lee commanded elements of the 2nd U.S. Cavalry, whose mission was to monitor Indians. He later commanded the Department of Texas at Fort Sam Houston in San Antonio before he resigned his commission to join Confederate forces. When the Civil War broke out, Camp Cooper was surrendered to the Confederates. The main Confederate presence along the frontier was the Frontier Regiment, a loosely organized unit that tried, with limited success, to protect settlers over an area far too large for the small force available. When it became apparent that Indian raids were becoming more successful, settlers gathered at civilian frontier "forts," such as Fort Davis in present-day Stephens County and Fort Owl Head in Shackelford County. These forts were little more than a cluster of houses, usually of picket construction, with protection afforded by the presence of armed civilians, which proved some deterrent to raiding Indians. The settlers, some of whom farmed along the Clear Fork, continued to see after their Longhorn cattle as best they could as the animals roamed at will across the unfenced ranges. These early ranchers continued what they called "cow hunts" to brand feral cattle roaming the river bottoms and surrounding hills.

When the war ended and the army returned to the frontier, a military post was established at the point where Highway 283 now crosses the Clear Fork. Fort Griffin, as the post was named, became one of the most famous of the Texas forts, its reputation based largely on the town that grew up along the river. Troops stationed there were African American soldiers whom the Indians dubbed "Buffalo Soldiers" because of the their hair, which to the Indians resembled that of the shaggy buffalo. The post was also where Col. Ranald Mackenzie mounted his expedition to the High Plains in 1874, during which he caught the Comanches and Kiowas in winter camp in Palo Duro Canyon and effectively ended their nomadic life by killing more than a thousand of their horses.

The town of Fort Griffin, or Griffin Flat as it was sometimes called, attracted a legion of frontier types that included gunmen, gamblers, prostitutes, petty thieves, and buffalo hunters along with the settlers. Names such as Pat Garrett, Doc Holliday, Wyatt Earp, Lottie Deno, Big Nose Kate, Hurricane Minnie, John Selman, and John Larn are part of the history of this colorful era. Among the ranchers were names such as Reynolds, Matthews, Irwin, Lynch, Cauble, Howsley, and others that are still found in the area today.

The town's popularity was a result of several factors. It was a major outfitting point for the buffalo hunters, who by 1879 had decimated the great southern herd, the result of a massive slaughter and incredible waste of a natural resource mistakenly thought to be limitless. Millions of dried flint hides, named for the hardness of the untanned skin, were brought in by the wagon load by hunters passing through Fort Griffin. Also during this time herds of cattle from South Texas were being driven on the nearby Western Cattle Trail. The trail drivers brought excitement to the town, especially to the saloons and merchants. But after the Indians were forced onto reservations in Oklahoma, the sale of goods to the hunters ceased, trail driving disappeared, and the post closed so that the soldiers could be moved further west to subdue the Indians there, life in the area settled into various agricultural pursuits, of which ranching was and still is the principal one.

The name of Chimney Creek Ranch, according to local historian Morris Ledbetter, derives from several stone chimneys that once stood along the creek that runs through the ranch. They were all that remained of settlers' cabins, but no one knows who constructed any of them. Noted Texas historian A. C. Greene, who has visited the site several times, has never seen even the remains of these chimneys.

Another local historian, Joan Farmer, believes that the chimney referred to in the name was part of the house built along the creek at Smith Station, a stopover on the route of the Butterfield Overland Mail, which crossed the ranch. The success of the operation, which ran from St. Louis, Missouri, to San Francisco, California, in the late 1850s, depended upon a series of well-organized and carefully located

BOTTOM: *Robert B. Waller*

relay stations to serve its passengers. Waterman Ormsby, a journalist who rode in the first coach going west, reported that the only housing at Smith Station on that opening run was a tent for the passengers and the family of the station operator, whose name was Smith, and a partially erected brush corral for the horses. Ormsby noted that the brush wall of the corral was chinked with mud. Supper that evening was some "cake cooked in the coals," dried beef, and some "clear coffee." Little other record exists of the stop. Later, a stone house was constructed along with a stone corral to hold horses. Nothing remains today of either. Apparently, the stones were sold during the Great Depression to the contractor building nearby Highway 6 and were crushed into gravel for the roadbed. In the 1950s when mesquite brush began to encroach on the grazing land, the owners arranged for two bulldozers to pull a large chain across the land to knock down the brush. This dragging obliterated any sign of the structures. A historical marker located on ranch property along Highway 6 stands as the lone monument to this historic effort at moving mail and passengers across the vast West. The effort ended as the Civil War loomed, and service by horse- or mule-drawn coach ceased, replaced later by the iron horse, the railroad.

In the 1870s a unique house was built on the property that would become Chimney Creek Ranch, but there is no record of who lived there. The house was square and constructed of limestone quarried near by. A bedroom and living room occupied one level, and up about five steps was another bedroom and a porch. The kitchen was in a separate structure. The most unusual feature of the house was a basement accessible from the rear, where the land slopes toward the creek. Local legend has it that the basement served as a place to hide horses from lurking Indians who frequently raided in the area, and that the house was used to cache arms for the Confederates during the Civil War. The generally acknowledged building date of the house in the 1870s, years after the war had ended, makes either of these stories unlikely, but the legends persist.

Ownership of the land passed through several hands in early years, but by 1882 the ranch was in the hands of Frank Eben Conrad, a well-known and successful merchant in Fort Griffin. Farmer states that Conrad and Charles Rath had the sutler's store that provisioned the military post and later moved to the town along the river. Conrad also later opened a store in Albany. He had done a booming business in food, clothing, lead, gun powder, rifles, knives, and other supplies for the buffalo hunters and trail hands, and he had also bought the buffalo hunters' hides.

After the military post closed, the hunters left to pursue other work, the frontier types moved on to other lawless towns further west, and the town of Fort Griffin languished. When the Texas Central Railroad laid its tracks through the area, the line went through Albany, not Fort Griffin, and the formerly bustling town died. Conrad closed his Fort Griffin store and concentrated his efforts in Albany. Apparently, Conrad divorced his first wife, who bore him one son, Frank B. Conrad, who married Ella Matthews, a girl of sixteen. Her family was prominent in the region and across the West. Her brother, J. A. "Bud" Matthews, had married Sallie Reynolds, who later authored *Interwoven*, a classic volume on life in the area. Miss Reynolds was the sister of several brothers who pioneered ranching and trail driving in several western states, including New Mexico, Colorado, Wyoming, North and South Dakota, and Arizona. They were peers with the legendary Charles Goodnight and Oliver Loving, who established several Texas ranches, the main one being the Long X at Kent, which is also included in this volume.

Conrad committed suicide on his birthday in 1892, apparently distraught over an unfortunate situation involving his first wife. Farmer records that he had believed she was guilty of adultery, but discovered his error and realized he had wronged her. Conrad left Ella with five children under the age of ten, all of whom had unfortunate lives. The daughter, Mary, died suddenly after playing in the snow. Joseph was killed on the ranch when his horse ran under a tree limb and crushed his skull. George committed suicide, and John died of a heart attack in 1945, the same year that Ella died at the age of eighty. Only Louis outlived his mother.

Ella continued to run the ranch after her husband's death and depended heavily on the advice of her brother, J. A. "Bud" Matthews, who owned Lambshead Ranch on the Clear Fork. It would later be the home of one of the most famous modern cattle barons, Watkins Reynolds Matthews. Mrs. Conrad granted permission for the Texas Central Railroad to lay tracks across her property and then constructed a set of cattle pens to serve as a shipping point for her own and her neighbors' cattle. She named it Bud Matthews, Texas, to honor her brother. In some years, more than 100,000 head of cattle were shipped from the site. By this time the ranch had grown to 14,000 acres.

In 1920 George Robert Davis bought the ranch from Mrs. Conrad's heirs. Davis, already established in ranching near Breckenridge in neighboring Stephens County to the east, had moved to the area from Kaufman, Texas. His sister Alice married Breckenridge Walker, the first white child born in the new settlement and for whom the town was named. Davis sold his Stephens County property to pay for the new ranch. He moved his family to the ranch, added on to the house, and later moved a small house to the site and joined the structures into one. He constructed a rock wall around the yard and built a small bunk house and a large barn. His family members still own the property, and the operation still uses the buildings he constructed.

Davis moved the ranch into the modern period. He enlarged the pens at Bud Matthews Switch, as it came to be called, when the Missouri-Kansas-Texas Railroad took over the line. He and his son Louie established the ranch as a producer of fine Hereford cattle and well-trained ranch horses. Among the cowboys who helped him were Clarence Holt, Lewis Burfiend, and Duncan Leech.

Successful ranchers often built homes in a nearby town or city so their families could enjoy the comforts of city life. Davis was no exception and erected his "prairie mansion" in Abilene, about thirty miles away. It was a three-floor structure with servants' quarters in the rear. He left the running of the ranch to Grady Smith (no relation to the operator of the Butterfield station on the ranch). Smith kept a busy schedule typical of cowboy work in that day, working seven days a week. The threat of screwworms kept him and other cowboys "prowling" through pastures on a regular basis, especially in warm months when infestation was at its worst. Smith's daughter Florene recalls that her father spent five days a week in the saddle looking after the cattle and worked weekends building and repairing fences or doing the other chores necessary to keep the ranch going.

By the 1930s the pens at the Bud Matthews Switch had fallen into disrepair, so Smith undertook the rebuilding, despite the effects of the Great Depression. He traded hogs raised on the ranch to Parker Sears, who ran the lumber yard in Albany, to help pay for the pine lumber used to rebuild the fences. He hauled the posts from Palo Pinto County to the east, where cedar growth has long supplied one of the favorite natural wood posts for such purposes. Bud Matthews Switch had a new lease on life and served neighboring Bluff Creek Ranch, Cook Ranch, Dawson Conway Ranch, and McComas Ranch.

One of the most interesting incidents or "wrecks" (a cowboy term for disaster) involved the shipment of a small herd of buffalo. Buffalo are very difficult to manage, but the cowboys finally succeeded in capturing them in the pens. Once the rail cars had been moved into place, the men began loading the animals into the car. The buffalo, distressed at being confined, began using their strong heads and horns

to break boards off the sides of the cars. The conductor, who had some ranch experience himself, encouraged the men to finish loading the animals before they tore up the rail cars. After the cowboys finished their work and the train pulled out, the buffalo settled down and arrived at their destination.

In the early 1940s, Smith was injured in a riding accident. As he was attempting to mount his horse, his foot slipped in the stirrup, and he pulled the horse over on himself. Smith left the ranch in 1945 because he had not fully recovered. He purchased a store in nearby Hamby that he operated until he retired in 1976. Because the store was on the highway between Abilene and the ranch, members of the Davis family often stopped by to visit. Smith died in 1986.

When G. R. Davis died, he left behind three children: Louie, Robbie, and Oma Frances. Today, Chimney Creek and other ranches that Davis purchased over the years—one in Throckmorton County and another near Midland—have been divided among the surviving heirs. Part is owned by Louie Bob Davis and the rest, including Chimney Creek, by Mary Frances "Chan" Driscoll. She lives in Midland and leases the ranch property to Waller Cattle Company. This absentee management began in 1957 when Driscoll's mother, Robbie, inherited the property and realized that she would be unable to operate it. She leased it to C. B., Robert, and Ruby Waller, doing business as Waller Cattle Company. Robert, the son of C. B., took over the ranch in 1973 after his father died and then took as a partner his son Robert, a graduate of Texas Tech University with a background in agriculture.

The Wallers have sought to improve the ranch by cross-fencing the pastures. The largest pasture was eleven sections, or 7,000 acres, when the Wallers took over the ranch. Another pasture that contained four sections has been subdivided into four pastures of 640 acres each. These are small compared to some ranches, but the Wallers feel that they have better control of their range and rotation grazing with this method, one that is growing in acceptance in contemporary ranching.

The Wallers do not raise horses but instead buy the Quarter Horse geldings they need for the work, usually keeping about six for the remuda of using horses. The cattle herd includes about 500 cows, 150 replacement heifers, and 40 bulls, about the optimum number for this ranch. The Wallers originally favored breeding Black Angus bulls to Hereford cows, and then phased out the Herefords. They then started crossbreeding the Angus-Hereford offspring, usually called Black Baldies, with Chianina and Maine Anjou bulls. The latter are black with some white, sometimes white stockings on the legs. The Wallers have also used some Charolais and Limousin bulls.

Since Chimney Creek Ranch is largely on a rolling plain, it lacks the draws necessary to provide enough tanks to water the stock. Although subsurface water is rare in this area, a well along the creek near the headquarters provides water for the family's personal use and is pumped to twenty-two troughs strategically placed in the different pastures so that cattle do not have to walk far to find water. The Wallers must check these regularly to ensure adequate water for the stock. This pipe network is essential to the ranch's operation.

The large open areas found on Chimney Creek and two more ranches to the east convinced area game conservationists, led by Lambshead's Watt Matthews, that antelope might find it suitable habitat. The mature antelope that were brought in and released did indeed thrive, but the young proved easy prey for coyotes. Since an antelope reacts to danger by fleeing rather than fighting, the females would abandon the young who were too weak to keep up. The surviving antelope have since died, and the experiment was deemed a failure.

Railroad service in the area was discontinued in 1967, and the rails were torn up in the early 1970s.

But the Bud Matthews Switch continues to be the shipping point for Chimney Creek Ranch. Mrs. Driscoll restored the pens in 1992 and placed a restored cattle car at the site to honor those who established the facility. She also secured a historical marker from the State of Texas for the site and had the Butterfield Overland Mail historical marker relocated to the site along Highway 6, where it is plainly visible to travelers.

Mrs. Driscoll has the ranch in a trust with a bank in Fort Worth and has built a small house in which she stays from time to time when she visits the ranch. She says, "This is my heaven on earth. Chimney Creek Ranch is a peaceful spot." Because she spent time there with her grandparents as a child, she reveres the ranch and its history, which, with its ties to Fort Griffin, the Butterfield Overland Mail, the Reynolds-Matthews family, and the progeny of G. R. Davis, is indeed a valuable part of the heritage that is Texas ranching.

Goodnight Ranch

FEW NAMES CONNECTED WITH RANCHING CONJURE UP IMAGES of the Old West like that of Charles Goodnight and the ranch he founded that still bears his name. The headquarters of Goodnight Ranch and Cattle Company is in the Armstrong County town of that name, which also includes land that once belonged to Goodnight and the famous JA Ranch in Palo Duro Canyon in the Texas Panhandle. Goodnight was instrumental in founding the JA and operated it for many years in conjunction with members of the Adair and Ritchie families, whose descendants still run the ranch to this day. The Goodnight Ranch, however, is a separate entity.

Goodnight's early life foreshadowed his later accomplishments. Born on a farm in Illinois, Goodnight moved to Milam County in Texas before he was ten years old. What little formal education he got came before this move. He worked wherever he could, once as a racehorse jockey. He farmed and drove ox teams for a time, and later, with his step-brother, Charles Sheek, the son of Mrs. Goodnight's third husband, he ranged cattle along the Brazos River. In 1857, just as the area of Palo Pinto County west of Fort Worth began to attract settlers, the two young men constructed a log cabin there for themselves and their parents. Goodnight eventually became acquainted with Oliver Loving, who ranched in the same area, and the two later took a herd of cattle to feed hungry miners in Colorado. In this endeavor Goodnight found his partner and his calling—raising cattle and trail driving.

Goodnight's ranching operation was not his only activity, however. He was forced to cope with the Indians, and he became a guide and scout for Capt. Jack Cureton's rangers. One of his most notable accomplishments was discovering the trail to the camp of Comanche chief Peta Nocona. Led by Cureton and Lawrence Sullivan "Sul" Ross, the rangers raided the camp and recovered Cynthia Ann Parker, who had been kidnapped as a girl decades earlier. She eventually married Peta Nocona and gave birth to the famous Comanche chief Quanah Parker, who later led his people into a new era of reservation life and became Goodnight's friend. Goodnight was also part of the famous Frontier Regiment serving the Confederate cause and became familiar with North Texas and the Panhandle.

Goodnight was to become one of the legendary cattle barons. With his older partner Loving and ranching giants such as George and W. D. Reynolds (whose exploits are described in the Long X Ranch chapter), Goodnight drove cattle over much of the West. To aid the trail driving efforts, Goodnight invented in 1866 the first chuck wagon, a basic tool of cowboy life that continues to serve as home away from home for many a cowboy. Goodnight and Loving are especially well known for their use of a route through West Texas into New Mexico and Colorado. The Goodnight-Loving Trail, which followed the Butterfield Overland Mail route from the Brazos River near Fort Belknap to the Pecos River and then up to its headwaters, was the primary route for driving breeding stock as well as herds to market into New Mexico and Colorado. It was on one of these drives in 1867 that Loving was fatally wounded in a fight with Indians. His body was taken back to his home in Parker County, Texas, for burial. Goodnight and Reynolds were instrumental in transporting the body home.

A less well known part of the Goodnight-Loving Trail is the northern portion that ran from Fort Sumner, New Mexico, the famous landmark associated later with Billy the Kid, to Las Vegas near Santa Fe, still following the waters of the Pecos River. There it joined the Santa Fe Trail, used by numerous traders making trips from the east to New Mexico, and then on to Raton Pass and east of the mountains into Colorado, to Trinidad, Pueblo, and on to Denver. Goodnight later rerouted part of the Santa Fe Trail by crossing the Gallinas Valley and the plains near Capulin Mountain before heading to Raton Pass (Richardson, 244). Goodnight also made other alterations to the route that led eventually to Cheyenne, Wyoming. Goodnight's dislike for a toll road over Raton Pass operated by Dick Wooton led him to devise another route through Trinchera Pass so that he could avoid it.

Like many other business tycoons, Goodnight was not always successful, but he always fought back. In fact, one of his most significant accomplishments, his founding of the JA Ranch in Palo Duro Canyon in 1876, was the result of his seeking to recover from a severe financial loss in Pueblo, Colorado. Among his less successful endeavors was his pioneering experimentation with crossbreeding buffalo and Angus cattle to produce cattalo, which proved to have little value for the industry. Goodnight also experimented with reintroducing wild game.

Goodnight's fame, of course, came from his ranching accomplishments, and he was one of the first five people inducted into the National Cowboy Hall of Fame in Oklahoma City. He also has some small claim to fame for producing a film of a buffalo hunt, the only one known to include Indians and wild buffalo. Only Sam Chance, the fictional rancher in Benjamin Capps' *Sam Chance*, comes close to representing the reality of Goodnight.

In 1887 Goodnight left his partnership with the Adair family (detailed in the chapter on the JA Ranch) and traded his interest in the JA for the Lazy F Ranch at Quitaque, which had been purchased for Cornelia Adair in 1882. He continued to serve as an adviser to Mrs. Adair until 1888. Financial hard-

ship forced him to sell half of the Lazy F to L. R. Moore of Kansas City, and he terminated his interest in the ranch in 1890. He had founded his own ranch in 1887 on 160 sections of land in Armstrong County (Anderson, "Goodnight, Charles," 245) and had established the headquarters near the newly established Fort Worth and Denver City Railroad. His two-story ranch house, featuring roofed porches and indoor bath facilities, was the first building constructed in the town, which was named after Goodnight (ibid.). Even though he was not formally educated, he believed in the value of education to the point that he founded a college, also named for him, in the town of Goodnight. It opened in 1898 with the help of Dr. Marshall McIlhaney, who served as the first president. Classes were held at the Goodnight Methodist Church. Goodnight donated 340 acres of land for the school's use and offered the school to the Methodist Church, which refused it. Goodnight then gave it to the Baptist Church, which saw the school prosper and enrollments rise to 175. In 1906 a three-story brick building was erected, and the faculty grew from the original two to six. In 1914 it became a junior college, but competition from West Texas State Normal College (now West Texas A&M) and Clarendon College proved too much, forcing the school to close in 1917. The facility was used by Buckner's Baptist Children's Home until 1920, when it was abandoned. The administration building burned in 1918, and Goodnight gave the property to the local school district (Reynolds, 244).

Goodnight married twice, and he was especially fond of his first wife, Molly Ann Dyer Goodnight. She was a frontier wife and the only woman to live on the JA Ranch for several years. She drove one of the wagons from Colorado when her husband, John and Cornelia Adair, and four cowboys took a hundred Durham bulls to Palo Duro Canyon to begin the expanded life of the JA. A native of Tennessee, she had come with her parents to Fort Belknap, Texas, near present-day Graham, in 1854, when she was fourteen years old. She met Goodnight at Fort Belknap, the home of the Second U.S. Dragoons during this pre-Civil War period. When she married Goodnight, he was ranching near Pueblo, Colorado, and three of her brothers worked for the ranch over the years. Goodnight's devotion to his wife was strong, and he even devised a sidesaddle with two horns so that she could rest both legs while riding. It was Molly who rescued the baby buffalo calves that became the herd for which her husband became famous. She died in 1926.

Goodnight's second wife, Corinne, was a young woman whose relationship with him is uncertain. She had nursed him back to health after an illness following Molly's death, and in 1927 he married her on his ninety-first birthday. They sold the ranch soon afterward, but he retained a life estate in it. Although he moved to Arizona for the milder climate in the last year of his life, his body was returned to Goodnight for burial in 1929. Thus ended one of the most colorful careers in the legendary lives of the Cattle Kings of the American West.

When Goodnight sold the ranch, J. R. Staley bought at least some of it and sold it to the Great Southern Life Insurance Company, which in 1938 sold to Mattie Hedgecoke 13,382 acres that included the original Goodnight headquarters. Within a year, Hedgecoke bought from Montie Ritchie of the JA Ranch the 42,000-acre Pleasant Ranch in Palo Duro Canyon, sixteen miles from the town of Goodnight. Hedgecoke's father, James Andrew Whittenburg, was an entrepreneur with a special interest in oil and gas. Among other accomplishments, Whittenburg founded a Panhandle town carrying his last name to provide a place for local oil workers to live. Later it merged with the town of Pantex to form Phillips, Texas, named for one of the major oil companies in the Panhandle. A grandson established the Amarillo *Times*, which he later merged with the Amarillo *Globe News*.

The ranch was operated in the post-war years by

Mattie Hedgecoke's son, J. A., with Holbert Brace as foreman. Also working on the ranch were the aunt and uncle of a teenager named Lee Palmer, who worked there during the summers and even on weekends before World War II broke out and curtailed travel due to lack of tires. Palmer went off to service during the war, and when he returned, he enrolled at North Texas State University. In the early 1960s, while coaching at Corpus Christi Baptist College, he was offered the job as foreman by Mr. Hedgecoke and realized a lifelong dream of running a ranch. Palmer says he always felt he was born "trying to rope a cow." He lived on various parts of the ranch during his tenure as foreman, which ended in 1988 when the family stopped operating the ranch.

When Palmer first became foreman, the operation included using the chuck wagon when the men were working cattle. One of the main reasons was that the Pleasant Ranch properties lay some sixteen miles from the headquarters, and the use of pickups and trailers had not yet become common.

The ranch had about two thousand head of horned Herefords, but in an effort to get crossbreed vigor, Palmer introduced Simmental bulls. He wanted the crossbred calves to retain the Hereford coloration as much as possible, and this cross proved quite satisfactory, although the cattle had a little more white than the Herefords. During the drought of the mid-1950s, they sold about half of the herd, including many of the old cows.

Palmer kept the bulls and cows in separate pastures except during breeding season. This was difficult at first because Pleasant Ranch was divided into only two pastures. Later Palmer cut the pastures into smaller tracts using barbed wire and steel posts, the latter because of the threat of fire. The cross-fencing made working the cattle and gathering the bulls more efficient.

Feed for the cattle was cottonseed cake shipped by rail to Goodnight for the headquarters herd or the town of Claude for the Pleasant Ranch. The feed was hauled by wagons pulled by mules or in a Dodge Power Wagon, the forerunner of the four-wheel-drive pickups that have revolutionized ranch transportation.

During Palmer's tenure, the horses used on the ranch were the offspring of Arabian stallions and utility mares, of which there were close to a hundred. Since screwworm was still a menace, the ranch hands spent a lot of time riding so they could find and doctor infested calves, and they produced some excellent ranch horses. Each of the seven cowboys had five or six horses in his string, depending on where he lived on the ranch and how much riding he had to do. When the crew needed to work at the Pleasant Ranch, they would bring a remuda of about thirty horses with them, and the only way they had to haul horses in Palmer's early days as foreman was in the back of a truck.

Although Arabians sometimes lack the innate cow sense of Quarter Horses, Palmer had excellent luck with this cross, particularly because one of the stallions seemed to imbue his offspring with the natural ability to work cattle. They also inherited the Arabian's legendary endurance, enabling them to cover the long distances required by roundups in the big pastures. Unfortunately, some of the horses had mean temperaments, which Palmer says came from the mares, which would "fight a man like a lion." If the horse were going to buck, however, it would do it in the morning and then work fine the rest of the day. The main stallion was gray, as were most of his offspring.

The cowboys lived on the top floor of the headquarters house that Goodnight built when he established the ranch. The shower was outside. Meals were cooked and served in the house. A bunkhouse on the Pleasant Ranch met their needs when working there. Saddles were mainly from local makers such as Bob Marrs, Carl Darr, Pete Borden, and a man named Sweitzer at the Matador Ranch. Palmer recalls that Darr's saddles were light but strong, and

CHARLES
AND
MARY ANN DYER
GOODNIGHT
TOGETHER THEY CONQUERED A NEW
LAND AND PERFORMED A DUTY TO
MAN AND GOD. HE WAS A TRAIL
BLAZER AND INDIAN SCOUT. SHE
WAS A QUIET HOME-LOVING
WOMAN. TOGETHER THEY BUILT A
HOME IN THE PALO DURO CANYON
IN 1876. THEY DEVELOPED THE
CATTLE INDUSTRY. THEY FATHERED
HIGHER EDUCATION AND CIVIC
ENTERPRISES.

TO THEM THE PANHANDLE PAYS
EVERENT AND GRATEFUL TRIBUT

Borden's were "horse high and bull tight," an expression that suggests they were heavy and strong. When working calves, the men usually roped and dragged them to the branding fire, but if Palmer had strong college boys working, he would have them simply wrestle the calves down.

Land management was one of Palmer's concerns, but the land was all but ideal cattle country, with the principal fodder being mesquite, buffalo, and sideoats grama. The main brush problems were mesquite and cedar, with the cedar being especially pernicious. Palmer says, "You can live with mesquite because it does not compete with the grass as much as cedar does. A mesquite tree has a deep root that will keep it going. Cedar roots are near the top of the ground and get the same water grass needs. Cedar will kill the grass around it." Chemical spray was used to try to curtail the brush, but with limited success, and the ranch was plagued with fires started by lightning. Palmer believes that, since the fires helped to clear the brush, they did more good than harm.

Although game once abounded in Palo Duro Canyon, when Palmer first went to the ranch few deer had survived encroaching civilization. However, some transplanted deer thrived, and there is now a healthy population.

Water was available in most pastures on the Pleasant Ranch from the Prairie Dog Town Fork of the Red River, which flows through the canyon. The water is heavily mineralized to the point of being gyppy. Pastures away from the river were watered by wells pumped by windmills into tanks made of cement or metal, later replaced by fiberglass because of the mineralized water's deteriorating effect on metal. The soil in many areas is sandy and will not hold water to form stock tanks. If problems occurred with the windmills or the wells went dry during droughts, the river usually had water in holes, even if it had ceased to flow.

In 1988 Darrell Cameron and his son Casey leased the ranch after the Hedgecoke family saw that needed improvements would cost a considerable amount of money. Darrell now lives in Amarillo, but Casey and his family live on the ranch and run about 1,500 head of cattle. The Camerons are experienced cattlemen, rodeo competitors, and steer ropers. Darrell dropped out of school in the seventh grade and went to work as a cowboy. He has been in the cattle business in one way or another ever since. He has worked as a buyer for packing plants, built feedlots, and run ranches. He has been well acquainted with members of the Hedgecoke family since his childhood.

Darrell's passion is steer roping, and it is also his grandson's. Clay, now a senior at West Texas A&M University in Canyon, was Rookie of the Year for the Professional Rodeo Cowboys Association (PRCA) in 1998. Darrell's son Casey, who learned cowboying at his father's heels, was also a successful professional roper before he injured a shoulder. But he still works as a cowboy and, along with two other men, runs the Goodnight Ranch as well as other parts of the Camerons' operation. Like all modern cowboys, they are more mobile than the men of Palmer's day and can do the work with fewer men.

The family passion for roping has influenced the ranching operation. The Camerons buy young horses and, with the help of their cowboys, teach them the skills for roping. Once trained, these horses can be sold for a good price to ropers looking for a horse ready to work. Darrell is clear on the kind of horse they look for—heavy-boned horses of medium build (what Darrell calls "low center of gravity" horses) with short cannon bones and enough speed to catch a steer running full speed. The desired weight range is 1,200 to 1,300 pounds, and the preferred height is 15.0 to 15.1 hands. The Camerons prefer red or blue roans. The optimum age to begin training is four or five years (that is, full grown), and during the next four years the men ride and rope off of these horses in the pasture and arena and really work the horses in the evenings and in the winter when there is less ranch work. The horses are taken to rodeos for one or two years to familiarize them with the crowd noise they might encounter at big rodeos like those in Cheyenne, Wyoming, and Pendleton, Oregon. Then the horses are ready for the circuit at about eight to ten years of age. The ranch tries to keep four or five such horses ready to go at any time.

The Camerons have definite ideas about the kind of cattle operation they want to run. They focus on British breeds and buy much of their stock from northern ranges—bulls from Canada and steers from Montana, North and South Dakota, and Nebraska. Darrell did buy 250 head of young Santa Cruz cows from King Ranch in South Texas and reports that they are doing well on the colder, drier Panhandle ranges. Angus bulls, the ranch's preferred breed, run with these cows. Some Charolais and Hereford bulls are used as well, but Brahman blood is avoided. The Camerons know that feedlot operators on the northern plains want large calves, and they strive to provide several thousand of the best possible animals each year. Darrell feels the Goodnight Ranch is rough and not well suited to calving heifers, so the ranch buys young cows to stock the range, also British breeds. They do not buy crossbred cows and do not keep heifers from the Angus-Hereford cross herds on their Texas ranches.

The Camerons also have a ranch at Wheeler, close to the Oklahoma border. Like the Goodnight Ranch, it is a cow-calf operation. They run yearlings on ranges in New Mexico, mainly in Union County, as well as in Wyoming and Nebraska. In recent years Darrell has sold three thousand to seven thousand yearlings each year to Caprock Industries.

The ranch begun by the archetypal cattleman Charles Goodnight still stands as a monument to his dream and perseverance. Today it is still being run by men who consider ranch life the only way to live. Like Goodnight, Darrell Cameron left his education early and followed the rancher's way of life. Goodnight, who worked hard all his life and eagerly shared his often meager resources with others, left a larger-than-life reputation. It is easy to see why Lee Palmer was also attracted to ranching, and why the Camerons still hold to a way of life that Goodnight loved.

JA Ranch

THE JA RANCH AND PALO DURO CANYON ARE INSEPARABLE in the minds of anyone familiar with ranching in the Texas Panhandle, and justly so. At one time the JA controlled 1,325,000 acres in Randall, Armstrong, Donley, Hall, Briscoe, and Swisher Counties, much of it in Palo Duro Canyon, and had more than 100,000 head of cattle along with a large number of horses and mules. It is the oldest privately owned cattle ranch in the Texas Panhandle, the home of large ranches, most notable being the long-defunct XIT with its 3,000,000 acres. The canyon itself is one of the dominant landmarks in the Panhandle.

The founder of the ranch was Charles Goodnight, considered by many to be the archetypal frontier cattleman. In 1876, ruined by drought and financial collapse, Goodnight lost his holdings in and around Pueblo, Colorado, and set out to reestablish himself. He knew of Palo Duro Canyon, which he probably first saw while riding on scouting duty with Texas Rangers during the Civil War. Later he and a crew of cowboys drove 1,600 head of Longhorn cattle from along the Canadian River in New Mexico south to the canyon, and on October 23, 1876, Goodnight rode down an old Comanche trail to the bottom and established a ranching operation there. The wagons were dismantled and carried down the steep, narrow trail on the backs of mules, and the cattle were driven down a few at a time. His cowboys drove out the buffalo and installed the herd cattle in their place. After Goodnight had settled the men and cattle for the winter, he returned to Colorado.

It was a fortunate stroke of fate that in Denver at that time was John Adair, an Irish financier interested in investing in the cattle business. Because Goodnight had the ranch but no capital, the meeting later produced one of the most important partnerships in ranching history. In the spring of 1877, Goodnight and his wife, Molly; John Adair and his wife, Cornelia Wadsworth Adair; four cowboys; one hundred Durham bulls; and four wagon loads of supplies reached the canyon after a twelve-day journey, two days of which were without water. The Adairs made the trip on horseback; Molly drove one of the wagons. Thus began an incredible venture that persists to this day, because descendants of Mrs. Adair still operate the ranch. During a two-week stay in the canyon, Adair and Goodnight struck their famous deal.

John Adair, a European aristocrat, owned vast estates in England and Ireland. His initial interest in the United States was in arranging loans at interest, so in 1866 he established a brokerage house in New York. At a dance in New York for Congressman J. C. Hughes, he met Cornelia Wadsworth Ritchie, the widow of Montgomery Ritchie, son of a prestigious Boston family. They married the next year. The couple spent time in both Ireland and America.

Adair developed an interest in the West when he heard of possible fortunes to be made in mining, and he moved his brokerage house to Denver, Colorado. He was somewhat familiar with life on the plains from a long journey in 1874 from Nebraska up the Platte River to Colorado. During that trip the Adairs went on a buffalo hunt with Goodnight, who spoke of the Palo Duro Canyon and its virtues as cattle country. On the hunt, however, Adair was seriously injured while chasing a buffalo. When his horse fell, Adair's weapon discharged, killing his horse, and Adair was thrown to the ground.

Adair later sought out Goodnight in order to go into the cattle business. Together they bought the Matador Ranch, which had belonged to a Scottish syndicate, and the T Anchor Ranch, the property of an English group. Adair and Goodnight entered a formal five-year partnership, with Goodnight owning one-third of the operation, and Adair two-thirds. Goodnight had to borrow his share of the purchase price from Adair, at interest. Goodnight received a salary of $2,500 a year for running the ranch. With the financial backing he needed, Goodnight began buying choice pieces of property in and around the canyon where water and grass were good. By doing so, he hoped to control all of the water and keep outsiders from intruding. He soon moved his headquarters twenty-five miles east of its original site and built a four-room house using cedar logs. He also built a bunkhouse, blacksmith shop, and several other structures. Being so far from outside support, the ranch had to supply most of its own needs.

The partnership proved so successful, with a profit of more than $500,000 in the first five years, that Goodnight and Adair extended their agreement for another five years, and Goodnight's salary was raised to $7,500 a year. The ranch had grown to 93,000 acres, and Goodnight was looking for more land to buy. He had already purchased the Lazy F Ranch at Quitaque for Cornelia, who was keenly interested in ranching herself.

John Adair died in 1885, but Goodnight continued to work for Cornelia. In 1887 she traded the 140,000-acre Lazy F and 20,000 head of cattle to Goodnight for his interest in the JA—336,00 acres of land, 48,000 cattle, a number of mules and horses, as well as equipment. She also wanted the rights to the JA brand, which used her husband's initials. Although afterward Cornelia lived primarily in Ireland, she nonetheless contributed to area causes and projects, particularly the Adair Hospital, the Episcopal Church, and a YMCA building in the Panhandle town of Clarendon.

After Goodnight's departure from the JA, the ranch relied on a series of managers, one of whom

JA

was James W. (Jack) Ritchie, Mrs. Adair's son from her first marriage. He served only briefly before returning to New York at his mother's request. Their socialite friends had been appalled that the young man was isolated at the ranch, though he later remarked that those years were among his most enjoyable. He proved a competent rancher and lived in a dugout while serving as foreman of the ranch's Tule division, where steers were ranged. In New York he negotiated a deal with the police department to buy horses from the JA herd. While serving in the Boer War, he became a major in the British cavalry, a significant accomplishment for an American citizen. His experience on the Texas Plains had proved an asset to moving men, animals, and equipment across the plains of Africa. An active sportsman, James also may have had a taste for gambling. This caused friction between James and Goodnight, who insisted that he had caught the young man gambling with the hands, a forbidden practice on the ranch, as was drinking alcohol.

From 1891 to 1910, Richard Walsh was manager. An Irish immigrant, he had been at the ranch for several years before he became manager. His innovative crossbreeding of Hereford and Angus cattle proved very successful in regions other than the humid coastal areas, where Brahmans proved to be more profitable because of their resistance to disease. The ranch had a strong commitment to Hereford cattle, even to the point of importing them directly from England. According to one story, when Mrs. Adair discovered that one of the foremen had stocked part of the ranch with spotted San Simone cattle, she promptly fired him. Mrs. Adair also had a strong preference for the color of her horses and had only bays, which are brown with a black mane and tail, and often black stockings.

Jack's son Montgomery Harrison Wadsworth "Montie" Ritchie came with his brother Dick to the ranch in 1931 at the age of twenty-one, no doubt lured by the tales their father told of the ranch in the Wild West. Dick died in an unfortunate accident on an extended boat trip out of Corpus Christi to a fishing ground. He slept too close to the boat motor and was poisoned by carbon monoxide.

Montie was determined to make it as a ranch hand but ran into trouble with Timothy Dwight Hobart, the manager, whose son hoped to run the ranch and saw Montie as a rival. Hobart was no rancher but he had successfully sold the nearby White Deer Land and Cattle Company. He hoped to do the same at the JA. The Hobarts played several tricks on Montie, including giving him only the wildest horses to ride. Never afraid of a challenge, Montie saw through their plan and set out to thwart it. He traveled to England to talk with the heirs, and when Montie returned to the ranch, he was armed with power of attorney. In 1935 he ousted the Hobarts and took over management of the ranch, which he ran until his daughter, Ninia Ritchie Bivins, assumed control in 1993.

Montie was an interesting man. He was born in Ashwell in Oakham County, England, on December 2, 1910. Since his mother was an American, he was also a citizen of the United States. He was educated in Switzerland and at Cambridge University, but his attraction to the ranch was strong and never wavered. His time of service as a Navy flyer during World War II took him away from ranching only temporarily. When he took over the ranch, it was in trouble from debt and drought, but through careful management, he paid off the debt, settled the estate business with other heirs, survived the drought, and reestablished the JA as a successful operation.

When Montie's sister, Gabrielle, asked that her share in the ranch be sold so that she could continue to live in England, he complied. One of Ritchie's strategies was to reduce the size of his direct responsibilities to about 100,000 acres. He organized the remaining 130,000 acres into eight different ranches for lease. He had learned the same lesson that the managers of the XIT Ranch had—even in Texas

something can be too big to manage well. He did, however, buy another ranch between Colorado Springs and Denver, where he spent much of his time while still managing the JA. Although Montie continued to rely on Hereford cattle, he also continued crossbreeding with Angus, even though he once said in an interview that the crossbreeding resulted in "mongrel" cattle, and he preferred to rely on the pure Hereford breed (McCoy, 119).

The ranch is now in the hands of Ninia Ritchie Bivins, Montie's only child. She has one son, Andrew. Following the pattern of John Adair, she formed a partnership to manage the ranch, choosing Jay O'Brien, a well-known Panhandle cattleman. In the past five years a major effort has been made to improve the ranch by rebuilding pens with metal pipe, reworking boundary and interior fences, improving the watering facilities, and rebuilding the cattle herd. The ranch runs a cow-calf and yearling operation. Advantages to running yearlings include the ability to adjust quickly to adverse range conditions. O'Brien is not as concerned with particular breeds of cattle as he is with cattle that produce efficiently.

Improvement of the horse herd has focused on producing quality horses for ranch work, not for the cutting arena, roping pen, or racetrack. O'Brien says the cowboys rope as little as possible to avoid injuring the stock and making them wild. A distinguishing characteristic of the horse operation is the easy method of training colts instead of bucking them out in order to break their spirit. The mares were bred from the ranch's own herd, but the current stallions are Señor Super Smoke, a bay descendant of the famous Quarter Horse foundation sire Doc Bar; and Freckles Tivio Bar, a gray descendant of Poco Tivio and Colonel Freckles. Their offspring are proving to be excellent ranch horses with lots of cow sense.

The cowboys on the ranch use typical cowboy tack. Saddles have swelled forks with rolled cantles, the lariats are nylon and average thirty feet in length, and the bridles have shallow port bits and split leather reins. In short, the ranch hands belong to the cowboy culture and show no influence of the South Texas vaquero tradition.

O'Brien brings considerable experience to the management of the ranch and, as Goodnight did, runs it as a partnership with the family owner. After working summers on a ranch near Roswell, New Mexico, while in high school, O'Brien attended Yale University. He entered the cattle business in 1967. He manages two other ranches, the Exell for other owners, and the Swamp, which he owns. His goal for the JA is clear: "to continue the work that Montie Ritchie did up until the early seventies in improving the assets and the herd. We have done it by working with dedicated, quality cowboys who could keep focus on the long-term goals." The intent is to have the ranch in good shape when Andrew Bivins takes over the ranch, so "he will have an easier job and a smoother transition." O'Brien says, "I look at ranching as a business. However, I feel blessed to have a business where I work so closely with the land, animals, and great folks. My success in the industry is related to luck, hard work, great associates, and the utilization of computers for record keeping." This emphasis on computers is evident in the ranch's extensive Web site.

In the process of running the ranch, O'Brien has become very aware of what he calls the "power of the JA history." He says, "People want to work for the JA because of the history, and people like doing business with the JA because of its history." He credits foreman Billy Hollowell with "making the JA a place that cowboys can be proud to work because it has a great team and history. . . . It is a pleasure to be able to be the steward of a beautiful ranch and work with cowboys who appreciate the land and livestock."

The family's commitment to the ranch remains strong. At the celebration of the ranch's one-hundredth birthday, Montie Ritchie summed up his view of the ranch: "Nobody or organization succeeds alone in this world, so the fact that we are able . . . to celebrate our one-hundredth birthday is due in large part to the wonderful, loyal men and women, leaders who worked for and with us, men of imagination, men of skill, men of courage, men who braved the elements day or night, men who took pride in their crafts, loved their horses, and understood their cattle and were eager to enhance the reputation of the JA and proud to be a part."

Ninia Ritchie Bivins continues this tradition and trusts that her son, Andrew, will share this commitment. With this kind of tradition, it appears that the Texas Panhandle will continue to have its flagship ranch, the JA, for a long time to come, and that it will maintain its heritage alongside the best that Texas has to offer.

Moorhouse Ranch

THE MOORHOUSE RANCH COMPANY HEADQUARTERS LIES SOUTH of Highway 82 on the eastern edge of King County, an area devoted to ranching. The county seat is Guthrie, a small ranch town west of the Moorhouse Ranch and adjacent to the headquarters of the 6666 Ranch. Benjamin, the seat of Knox County, is about fifteen miles east of the ranch. This rugged stretch of rangeland is well suited for grazing cattle and supporting wild game. The most distinguishable natural landmark in the area is Buzzard Peak, a brushy promontory south of the ranch on property owned by the 6666 Ranch, with which the Moorhouse shares a fence on the south and west. Little Croton Creek runs through the property. Also in King County are some of the holdings of the Spike Box Ranch, the Masterson JY Ranch, and the Pitchfork Ranch.

The story of the Moorhouse Ranch Company begins in 1907, when Ed, Percy, and Charlie Moorhouse bought 33,000 acres of the R. B. Masterson Ranch. After only two years, they were unable to make the payments and lost it. Ed then bought the Ross Ranch, but ruined financially by the drought of 1917–1918, he sold out. He had also bought the Thorman Ranch, the current headquarters ranch, but lost it as well during these difficult economic times. Ed and his wife had eight children: Frank, Walter, Myrtie, Coleman, Zell, Bowlden, J. C. (Togo), and Mabel. Not long after he lost the Thorman Ranch, Ed died.

Despite misgivings about moving to the area, Ed's wife had been pleased to hear of what sounded like a delightful spot known as Fern

Mountain. Instead of a mountain covered with green, sweet-smelling fern, however, she discovered a small hill covered with rock and cedar brush, which had been named in honor of Fern Masterson, the daughter of one of the area's prominent ranching families.

Togo, the founder of what is today called Moorhouse Ranch Company, was born in Kaufman County and was one of the younger boys. He was born during the Russo-Japanese War, and one of his uncles came up with his nickname, after the Japanese hero Admiral Togo Heihachiro (McCoy, 95). The family moved to Benjamin when he was two years old. After his father's death, Togo lived with his mother in Benjamin but became intent on becoming a cowboy and rancher. After he graduated from high school, he attended John Tarleton College in Stephenville, Texas, for one year. Tom Moorhouse says his father worried that if he stayed in college, there wouldn't be any cows left to punch when he got out, and he wanted to be a cowboy. He went to work as a cowboy for the McFaddin Ranch, a job he really enjoyed. He worked there about seven years and along the way formed a ranching partnership with his brother and Wayne Dolan, the manager of the McFaddin Ranch. Togo also began working toward being an independent rancher.

The group leased a ranch in Stonewall County and bought some steers to graze on it. They then put together a crew to drive some cows and calves to the ranch. They drove across open country, and when they came to Big Croton Creek, they had to use shovels to knock off the edges of the bank to get the chuck wagon across. When they reached the ranch, they put the cows and calves in a trap, gathered the steers, and took them to the railroad at Aspermont to ship to market. The men returned to the ranch and scattered the cows and calves. Well aware of the risk he was taking, Togo kept his job at the McFaddin Ranch for a while to keep a monthly wage coming in. It likely saved him during the Depression, when banks called in many of their loans.

Later he and his brother Bowlden bought the Thorman Ranch, which their father and his father had lost years before, and it has been the headquarters ranch ever since. In 1944 Togo convinced his wife, Lucille, to move to the ranch to help him feed the cattle that winter. They ended up staying for twenty-five years. When their youngest son, Tom, finished college and moved back to the ranch, they returned to town. By then Togo had become known as one of the modern cattle barons, even though he had never been lucky enough to have oil to help his operation, and did not inherit land. He died in 1995 at age ninety.

The ranch now owns the original Thorman property and another ranch of approximately equal size adjoining it. The leased properties include another 70,000 acres near Big Lake, 22,000 acres in Hall County near Estelline (part of the old Milliron Ranch), 25,000 acres in Stonewall County, the Alexander Ranch of 10,000 acres across the Wichita River, and 20,000 acres north of Benjamin. The ranch brands an AC on the left hip.

The ranch is managed by brothers Tom and John. John handles the business dealings and the farming; Tom handles the cowboy work. Tom is quick to admit that his first love is being a cowboy, and that he may spend too much time cowboying and too little time managing. But he so loves the work that he has trouble doing it any other way. All of his brothers—Ed, John, and Bob—graduated from Texas Tech University, but after Tom tried Tech, he felt it was too big and had "too much asphalt" to hold him. He transferred and earned a degree from Sul Ross State University, but he never was interested in anything but cowboying and worked on a ranch near Alpine while in college. Tom readily asserts that he is indeed doing the work for which he has yearned all his life. One indication of his success was his being named Top Hand at the ranch rodeo in Wichita Falls in 1981. He and Becky have five children: Jed, Brad, Jody, Seth, and Gage.

Tom and his four brothers grew up along with

the sons of Windy Jones, the man who helped with the Stonewall County Ranch. For several years they had a crew of nine cowboys: two men and seven boys. His father first had the boys ride double with them, and later they rode their own horses, often bareback. Tom has fond memories of the great rewards of working cattle with that group and learning the cattle business from the back of a horse.

One brother who no longer lives on the ranch but is deeply involved in cowboy work is also a photographer/artist. Bob Moorhouse is manager of the Pitchfork Ranch, which has its headquarters in the same county and is one of the best-known ranches in the world. An excellent cowboy and manager, Bob is also an outstanding photographer, particularly of ranch and cowboy life.

The house at the headquarters sits beside a large stock tank, practically a lake, that provides water for the house and the pens at the headquarters. In fact, since there are no wells, all stock water in the area is from tanks. Tom has great affection for this unique house that he and his brothers grew up in. The original part of the house, one room made of stone, is the oldest permanent structure in King County. The interior walls were once plastered and then covered with wood paneling, but Tom and Becky stripped off the paneling, chipped the plaster away, sandblasted the walls, and restored the interior walls to their original appearance. The kitchen was originally a small house by a spring about two miles from the headquarters. It was put on a skid and pulled by mules to the headquarters house, where it was joined to it with a breezeway. Later, the breezeway became the bathroom. The porches that were added later became rooms. Adjacent to the house is the old bunkhouse, which has been remodeled into an office. Tom remembers that as he and his brothers grew up, they moved from the house to the bunkhouse, an important transition in growing up as cowboys.

The area vegetation includes a considerable amount of cedar, especially on the western edge of the ranch, as well as mesquite and some prickly pear, but the brush is not as thick as that found further east. Care of the land includes the use of controlled burns to retard the growth of brush and cactus. The Moorhouses would like to burn more than they do, but there are only a few days from late January through March when the wind, temperature, and humidity are right for the burns. If a burn is done under less than optimum conditions, it does not do a thorough job of retarding the growth. To help with the controlled burns, the ranch has adopted the use of metal posts for any repairs or new barbed wire fencing.

The Moorhouses have long been known for quality horseflesh. One of Togo's brothers bought a Yellow Wolf stallion from the W. T. Waggoner Ranch and gave one of his colts to Togo to use as a stud horse. He was a dun horse with a dorsal stripe, a trait he passed on to many of his offspring, and Tom says he had lots of "bottom," or stamina. When the stallion died, the ranch replaced him with another Waggoner stud and has since brought other stallions into the breeding program as well. The ranch has about twenty-five broodmares and strives to produce high-quality horses with lots of natural cow sense, which are then ridden as much as possible. Hauling is kept to a minimum and used only for transporting the horses from one ranch to another. Tom says that when he hires a hand, he is as much concerned about the applicant's horsemanship as he is with other skills such as roping and working cattle.

In the old days the ranch broke horses by roping a front foot, or forefooting them, and bucking them out. Now they use gentler methods. When the time comes to work with the two-year-olds, all of those to be put into the ranch remuda go through training in the same sequence, and after a week it is typical for the hands to be riding these green-broke horses with very little difficulty. The horses are then parceled out among the hands for continued riding and training in cattle work. The ranch knows that these horses are worth more money well trained, so they are ridden and used to work cattle often, particularly for roping and dragging calves at branding time. Tom remembers that his father had an old calf cradle, which he discarded once the boys got old enough to flank calves. One of the basic rules of horse handling on the Moorhouse Ranch is that if a horse is in the remuda and is three years old, he has dragged calves for branding.

Saddles used on the ranch come from various sources. Tom remembers that in the early days most of the saddles came from Carl Darr, the Oliver brothers of Vernon, and H. H. Sweitzer of the Matador Ranch. However, the number of saddle makers has grown to the point that good saddles are readily available from many different companies.

The Moorhouse ranch is particular about the kind of cattle raised. Early in the operation the ranch adopted the Hereford breed, one that proved well suited to North Texas ranges. For calving ease the ranch used Black Angus bulls (a breed characterized by low calf birth weights) on the first-calf heifers, only to discover that at market time these calves, which would ordinarily be small because of the young age of their mothers, outweighed the Hereford calves. Consequently, they adopted the Black Baldy as their optimal calf for sale and kept heifers from this cross as well. After several years most of the cows on the ranch were Black Baldy. At that point they bought Charolais bulls for crossbreeding. The ranch still maintains a herd of about five hundred Hereford cows and their best Angus bulls on the Hall County ranch, and they purchase these from outside sources in order to keep their three-breed cross viable.

Tom's affection for ranch life remains strong, and he has extremely high regard for ranch women. He credits his mother and grandmother with the ability to be incredibly strong in the face of adversity and hardship in the early days of ranching, particularly in this part of the state. His mother came to that one-room stone house, where over the years, as it and the family grew, she raised four sons in the typical fashion of that day—for many years going without electricity, running water, and the other conveniences that town folk could expect. The ranch did not have electricity until about the time of World War II. Tom is equally complimentary of this generation of women, particularly of his first wife, Sue, who died, and his wife, Becky, who ably manages their household and also home-schools their youngest child, Gage. She also has begun making items out of rawhide, which has been little used in artistic work in this part of cattle country.

One tradition that fits nicely into the ranch's economics is the continued use of the chuck wagon. Many ranches have abandoned it in favor of hauling the riders and horses in pickups. But the holdings of the Moorhouse Ranch are scattered, so Tom still rolls out the chuck wagon, and the remuda moves right along with it as the crew works its way across a ranch. Tom sometimes wonders if the wagon is really a money saver, or if it's just that he really likes that kind of life, which captures the spirit of the open-range, trail-driving cowboy.

Leased hunting is one form of supplemental income for the ranch, and on the home ranch presents little difficulty. Problems can emerge, however, on the leased ranches. The lessee has to pay more for the hunting rights, but if he leases only the grazing, he must put up with hunters whose arrangement is with the landowner. It makes better sense to control all of the rights to the property, so when possible the Moorhouses lease both the grazing and the hunting rights and sublease the hunting rights themselves in order to have some control over who is on the property.

Tom has great fondness for his parents and is grateful for the monetary legacy of the ranches. But the more important legacy was the lesson he learned from them, that "if you want something, you have

The late Togo Moorhouse

to work for it, and if you're going to run a business, you have to think. They also insisted that if you are going to get by in this world, you have to be honest and have integrity."

Although the ranch controls thousands of acres, both deeded and leased, Tom does not consider the ranch to be one of the "big" operations. They have no oil revenues to supplement the income and must ranch economically to be successful. But the Moorhouse family is one that relishes the traditions and values of ranch life, and their careful practices in breeding cattle and breeding and training horses, as well as their conservation of the land, are marks of a proud heritage of ranching.

Nail Ranch

LYING NEAR THE NORTHERN EDGE OF SHACKELFORD COUNTY is the ranch of J. H. Nail, Jr., a 64,000-acre spread whose history goes back to the earliest days of Anglo settlement in this rolling, sometimes arid region. Comanche Indians once controlled this entire area, a history aptly documented in Rupert N. Richardson's *The Comanche Barrier to South Plains Settlement*. The Nail is today the largest contiguous ranch operating in an area that includes other large, historic ranches such as the Matthews, Green, and Caldwell.

The Nail family made a significant contribution to this region and others. The first of the family to come to the area was John Nail, a Tennessean who migrated to Texas in the 1830s and settled at Wolfe's Mill (later called Wolfe City) in Hunt County. He was killed by outlaws whom he was chasing with a posse. His son William, nineteen at the time, assumed the role as head of the household. He later married and fathered ten children, two of whom died in infancy. After William died from a fall from a horse in 1875, James Henry, the oldest son, took over the family business, which came to include cattle ranching and banking. The operation prospered under his management.

By the late 1870s and early 1880s, grazing land in Texas had become crowded by standards of the day. When the federal government offered leases on vast grasslands in Indian Territory, Nail and other Texas ranchers such as Dan Waggoner and Burk Burnett agreed to lease land from the Indians. It was on this grazing land that President Theodore Roosevelt joined some of the cattle barons on hunting expeditions.

In order to oversee the ranching operation, Nail opened a business office in 1885 in Dennison, just south of the Red River. He invited his younger brother, William Chapman, better known as Buck, to be his manager. The younger Nail, only sixteen, became widely known as an astute judge of livestock. Soon another brother, Dick, joined the business. Despite the difficult economic and climatic conditions of 1886–1888, the Nails' ranching operation flourished.

Bob Green, an Albany rancher familiar with the history of the Nail family, recalls an amusing incident involving Buck and Dick, who were riding a train held up by bandits. As the robbers were forcing the passengers to hand over their valuables, Dick handed Buck some money, saying, "Buck, here's that fifty dollars I owe you." Buck would later laugh about the incident, but the danger was quite real.

After Buck and Jim decided to divide their business ventures, Buck began working for a cattle operator named Atwood Reisner, who had extensive holdings in Indian Territory. Buck learned much from Reisner, who later hit a streak of bad luck and went broke, not an uncommon occurrence for land and cattle speculators. Reisner introduced Nail to George and W. D. Reynolds (founders of the Long X Ranch, also included in this volume), who convinced Nail to visit Northwest Texas and consider ranching there. The rest of the Nail story is associated with this part of the state.

Buck Nail liked the area and had some of his men in Indian Territory bring part of his herd to the Zugg Ranch, which he leased. Now known as the Akers Ranch, it lies between Abilene and Albany. Later he purchased land adjacent to the Zugg to expand his operation, and he invited Merrick Davis, the son of his sister Adeline Rebecca, to come to work on the ranch. Davis soon came to relish the life, and he later implored his mother, a successful hotel operator in Paris, Texas, to buy him a ranch west of his uncle's spread. Today, both the Buck Nail and the Merrick Davis ranches belong to members of the Matthews family but continue to bear the names of these early owners.

By the late 1880s Buck Nail's brother Jim, his sister Missouri Matilda, and her husband, W. I. Cook, came to the area to see about moving there themselves. The newcomers liked what they saw and bought the Holstein Ranch and Dobbs Farm north of Albany. Later, Jim Nail sold his part of this property to the Cooks, and that ranch came to be known by their name. Rich oil reserves were later discovered on the Cook Ranch, and the oil field is one of the most productive in the United States. A portion of the profits benefited Cook Children's Hospital in Fort Worth.

Jim later purchased a one-hundred-section stretch of rolling grassland cut with draws and creeks north of the Cook Ranch. This land, which had formerly belonged to the Monroe Cattle Company (in which the Reynolds brothers and J. A. Matthews figured prominently), became the Nail Ranch. Jim retained the original headquarters on Foyle Creek and settled in to become one of the area's most prosperous ranchers. The Nail's northern boundary fence roughly marks the route of the trail made famous by Col. Ranald Mackenzie, the Indian fighter who led cavalry troops on numerous forays to the High Plains before he caught the Comanches and Kiowas in winter camp in Palo Duro Canyon in 1874. There he burned their food supplies and shelters, and slaughtered their horse herd, ending the hold of these nomadic Indians on the region. Not far to the north of the Nail property stands the site of Camp Cooper, the home of Col. Robert E. Lee before the Civil War, and also the site of Fort Griffin, a post-Civil War fort and frontier town by the same name with a reputation for being rowdy and dangerous.

After Nail's death, his son James Henry, Jr., also known as Jim, assumed control and promoted the ranch to a new level of prosperity. He concentrated on raising Hereford cattle and quality Quarter

Nail

George Peacock, foreman and day-to-day manager

Horses. He also raised greyhounds, with which he loved to chase coyotes. Jim had two sons, Ronnie and Larry.

Jim's operation of the ranch from the 1930s through the early 1980s was in the classic tradition of ranch life. In the earliest days, about a dozen bachelor cowboys lived in the bunkhouse, and married cowboys lived in three camps placed across the ranch. Roundups were typical of the period on a ranch with lots of land, but still centered near the headquarters. The cowboys ate their meals in the cookhouse at the headquarters and then rode their horses to the pasture to be worked. This sometimes required riding ten to fifteen miles, which meant they had to leave headquarters by 4:00 A.M. to get to the pasture by sunrise. After gathering the cattle, the cowboys would eat the lunch brought to them or lope their horses back to headquarters to eat, ride back to work the cattle, and ride home that afternoon. If cattle needed to be moved, cowboys on horses drove them to the appropriate pasture.

Horses got a lot of work in the course of other chores as well. Ranch hands spent long hours "prowling" for cattle infected with screwworms, and checking each tank for cattle mired in the mud. Bogged animals had to be pulled from the mud with a rope. One end would be looped around the animal's horns or neck, and the other tied to the saddle horn. It is a fact of cowboy life that stock saved in this manner from a painful death almost always try to repay the kindness with a valiant effort to gore the man and his mount.

Taking cattle to market necessitated driving them to the railroad corral in Albany, a distance of almost ten miles. The trip could be made in one day. Tunny Hollingsworth, who worked for the Nail for more than thirty years before he retired in the late 1980s, remembers loading cattle and having to dodge their kicks while chasing them up the chute and into the railroad cars. The trick then was to place a bar across the door before the stock decided to bolt. A man not quick enough to get the bar in place was in danger of being trampled by a miniature stampede.

After World War II, shipping practices changed, and trail drives were abandoned. The first trucks were small and could haul only a dozen or so animals at a time to the railroad corral. The later cattle vans could haul nearly a hundred head, which were taken to Fort Worth or other markets. Truck drivers, latter-day trail drivers, do not have the romance associated with their work that the old-time drovers had, but they nonetheless perform the same work.

George Peacock went to work as a cowboy on the Nail in 1965 and later became foreman. When Mr. Nail died in 1983, Peacock became manager. The ranch was put into a trust with a Fort Worth bank, which handles the ranch's business dealings. A bank officer directs the general management of the ranch through Peacock, the on-site manager.

Long known for its quality Hereford cattle, the ranch has in recent years begun crossbreeding, especially with Black Angus and Limousin. The ranch keeps the bulls separated from the cows for most of the year and runs them together for only ninety to one hundred days to assure that calves are a uniform size at shipping time. Calving is usually completed by January 1.

Pastures include several thousand acres each, enough range to support about 150 head, a number the men can readily work in one day. There are about a dozen cow-calf herds on the ranch, so branding requires twelve days of work scheduled around the bad weather typical of January and February. The three cowboys who live in camps on the ranch are joined by cowboys from the adjoining Lambshead Ranch as well as day workers to form a branding crew of a dozen or so. The Nail crew then return the favor on the helpers' home ranches, a time-honored cow-country tradition called "swapping help." The men haul their horses to the pasture to be worked, unload the horses, and round up the cattle. The next step is to separate the cows from the calves. At noon, the crew rides in pickup trucks to the cookhouse for lunch. After lunch, the men complete the work and return to their home ranches to feed stock, check sick animals, and wrangle the horses for the next day.

As shipping time approaches, the cattle are pastured separately, with steer calves and their mothers kept apart from the heifer calves and their mothers. As July nears, the men bring the steer calves and their mothers closer to the shipping pens near the headquarters and eventually put them in the several-hundred-acre shipping pasture. Used only at shipping time, this pasture offers lush forage to put the last few pounds on the calves. On shipping day, the cowboys gather the herd and then separate, sort, weigh, and load the calves onto vans for shipment to pasture before going to the feedlot for finishing. This work is usually done by noon. After lunch, the men pregnancy test and cull the cows. Those to be kept on the ranch are driven back to pastures. The men cull the cows that are ten years old, as well as those with injuries or infirmities. Each cow carries a brand that indicates the year of her birth, and when her decade is over, she is sent to market.

About a week later the cowboys gather the heifer calves and mothers. Some of the heifers are sold. Others, already year-branded as young calves, become part of the breeding herd and receive the ranch brand at this time. In a year or so these young animals will be bred to Longhorn bulls to begin their years of reproduction. Using the slender Longhorn decreases calving problems in young cows.

The Nail Ranch's land-management practices include applying chemicals, especially to control prickly pear, and controlled burns to keep mesquite and other brush under control. The ranch also uses bulldozers to control brush. Keeping adequate forage on the range for the cattle is a primary concern, and uncontrolled brush will take all the moisture and keep the grass from thriving. The ranch also uses some limited "cell" grazing, a pattern of high-intensity grazing with extended periods of rest. Most of the pastures are still as large as they were when the ranch was first cross-fenced.

The many miles of roads on the ranch also require care, especially after wet weather. Since there is extensive oil production on the ranch, the roads are used daily by workers checking wells, heavy service trucks, and drilling rigs, which can turn a gravel road into a rocky quagmire. Consequently, the crew often has to interrupt cattle work and tend to road repairs using the two dump trucks, a front-end loader, and a grader. Maintaining the roads is important for preventing damage to the stock trailers used to haul horses or the trailer truck used for hauling cattle, both of which are necessary for an effectively functioning ranch.

Many people associate the windmill with ranching in West Texas, but the Albany area is an exception. Few spots in the area have subsurface aquifers into which wells can be drilled. In fact, most of the subsurface water here is brine associated with rich Mississippi Sand oil production. On the Nail Ranch, however, stands one of the largest windmills in Texas, its fan a whopping thirty feet in diameter. Electric pumps now provide this service, but originally the windmill pumped water from the Clear Fork of the Brazos River, which runs along part of the ranch's border, into a holding tank on a hill. From there water flows through an underground pipeline to earthen tanks on various parts of the ranch. Much of the ranch's water is supplied by an improved network of earthen tanks to catch and store surface runoff, the common method of watering stock in the area.

Although Jim Nail did not allow commercial hunting, it is now a major source of revenue for the ranch, which is known for trophy whitetail deer as well as turkeys, quail, and feral hogs, some crossed with Russian boars. Craig Winters is the game manager, and his duties include monitoring the game and food supplies, conducting game counts, and guiding hunters. Most of the hunts are guided three-day outings. Hunters sleep in the old bunkhouse, renovated for this purpose, and eat in the cookhouse. George Peacock's wife, Sue, one of the best cooks in the region, prepares the food. The ranch also offers a wilderness adventure hunt, with hunters riding horses to the camp and living "off the wagon" in frontier fashion.

The Nail Ranch rodeo team is one of the most successful in the United States. They have won the national title on two occasions, the most recent in Amarillo in 1996. The cowboys on the ranch also frequently ride in the *Fort Griffin Fandangle*, an outdoor drama depicting the settlement of the region in story and song.

The Nail Ranch has a proud history that goes back to earliest days of Anglo habitation in the region. The Reynolds-Matthews partnership may have originally established the ranch tradition here, but Jim Nail and his son, Jim, Jr., brought it to full fruition on this quality cow-calf operation in prime ranch country.

Renderbrook Spade Ranch

BARBED WIRE PLAYED AN IMPORTANT ROLE IN MAKING ranching in Texas what it is today, and Renderbrook Spade Ranch owes its very existence to it. The Ellwood family purchased large tracts of land in the region with money made from the sale of barbed wire, thus parlaying their fortune into one of the most significant ranches in the world. Descendants of the founder still control the ranch today.

Renderbrook Spade Ranch lies south of Colorado City between Abilene and Midland/Odessa. In this usually dry part of the state, few streams run year-round, but the Colorado River runs through part of the ranch. A major spring, discovered in 1872 by a cavalry patrol led by Capt. Joseph Rendlebrock out of Fort Concho, located at present-day San Angelo, continues to be important to the ranch. Although it is not the only spring in the region, it is nevertheless a strong one that long attracted travelers. Indians camped there, and Texas Rangers also found the spot a welcome stopover during the fence-cutting wars of 1883–1884. The spring would eventually be the magnet around which the ranch developed. The name of its discoverer was eventually corrupted to *Renderbrook*, the name associated with the ranch.

After the threat of hostile Indians lessened in the 1870s, Anglos began settling in the area. By 1878 J. Taylor Barr had built a dugout at the spring, and he later operated a livery stable. Before he died from the smallpox epidemic that swept through some years later, he sold his property around the spring to D. H. and J. W. Snyder, early stockmen who traded horses and drove cattle to New Mexico, Colorado, Wyoming, and Kansas. At the spring they built what Steve Kelton, in his centennial study of the ranch, called a "substantial headquarters," known as the "White House," as well as a bunkhouse (24). They began fencing the property to keep their Shorthorn cattle, which they were carefully crossbreeding with the native Longhorn stock, from mixing with strays. The disastrous drought on ranges across the West prompted the Snyders to sell their herds in 1887 to the XIT Ranch, which was at that time just getting established in the Panhandle.

The man who finally launched Renderbrook into what it eventually became had his beginnings far away from arid West Texas. A De Kalb, Illinois, merchant who specialized in hardware, Isaac Leonard Ellwood had a successful business selling merchandise and breeding quality draft horses with Percherons imported from France and Clydesdales from Scotland. In 1874, seeing the market potential in barbed wire, he formed a partnership with Joseph Farwell Glidden, the man who had pioneered the most practical design of the wire fencing. Over the following decades, Ellwood spent much of his time, energy, and resources in the struggle over ownership of the patent and manufacturing rights for barbed wire. His dealings with Jacob Haish and others in that trade war are the stuff of which business legends are made, with Ellwood's tenacity and business sense bringing his partnership out on top in that struggle.

Ellwood's association with John Warren Gates, an entrepreneurial salesman, is a colorful chapter in that story. Gates knew the market potential in Texas for barbed wire was enormous, but sales were slow, so he conceived a remarkable publicity stunt. To prove that it could hold cattle, he erected a barbed wire corral in downtown San Antonio and put wild Longhorn cattle in it. With this ingenious ploy, Gates saw sales boom, and he and Ellwood made a fortune from the rush for the wire that changed Texas ranges forever. Although the two men would be in conflict for some years over rights to the wire, they eventually formed a partnership, a move that proved once again Ellwood's business acumen. They formed the Columbia Patent Company, which was later converted into the American Steel and Wire Company and eventually added to John Pierpont Morgan's U.S. Steel Company.

Although deeply involved in his mercantile business, horse breeding operation, and the barbed wire controversy, Ellwood was also interested in Texas ranching. In the mid-1880s, he and his oldest son, W. L., came to Texas to market barbed wire and see the ranch lands for themselves. Legend has it that they paid a million dollars and a stud horse for the 130,000 acres that became Renderbrook Spade Ranch. Considering, however, that property in that part of Texas was selling for a dollar an acre at most, and more commonly as little as twenty cents an acre, it is doubtful the price was that high.

The Ellwoods' search for cattle to stock their newly purchased range ended with their purchase of eight hundred Shorthorn-cross cattle from J. F. "Spade" Evans near Clarendon, Texas. These cattle were already marked with what would become the ranch's legendary brand, the outline of a shovel lying horizontally. Ellwood decided to keep the mark and registered it in Mitchell, Coke, and Sterling Counties.

The family next faced a new challenge. Owning property and cattle does not make a successful ranch. The Ellwoods needed someone to run the operation, and the man they hired, David Nathan "Uncle Dick" Arnett, proved extraordinarily important in the ranch's development. A veteran of the Confederate cause, a Texas Ranger, an Indian fighter, and a cattle driver, Arnett left his own ranch in Mitchell County in his son's charge and went to work for the Ellwoods.

Hearing of some other property for sale by the Snyders, largely in Lamb and Hockley Counties to the north of their first range and west of Lubbock,

the Ellwoods traveled to the area and bought 128,000 acres. This property differed noticeably from the land near Renderbrook Spring. This was on the High Plains, and it turned out to be some of the most productive farmland in the United States. The deep, rich, sandy soil was made even more valuable with the discovery of the region's extremely strong aquifer, the Ogallala Formation. Also proving to be of great value was the foreman who came with the property, Frank Norfleet.

In the late 1890s, the ranch continued to expand with the Ellwoods' purchase of seven sections, known as the 49 Pasture, in Borden County west of the current town of Snyder. This range served as a stopover on cattle drives between the southern and northern properties, a ten-day trip. One advantage of the property is that it is north of the tick line, that point at which cold weather kills the ticks that were later found to cause Texas or tick fever. At first the southern ranch was a cow-calf operation, while the northern property was used to fatten steers until they were three or four years old. After fattening there, the steers were sold to other stockmen who would eventually sell them for slaughter. Shipment was usually by rail from the town of Bovina, but later pens were constructed on nearby Yellowhouse Ranch.

By this time Ellwood was tiring of the barbed wire trade and sold his part of the business, a wise move that produced a significant amount of capital. Ellwood invested his profit in ranch land, adding tracts that were next to or near his present holdings until he had put together 395,000 acres.

In the early 1900s, the ranch was fortunate to have a series of good managers. When Uncle Dick Arnett retired in 1912, his son, Dick, Jr., who had been hired as wrangler when his father came to the ranch, stepped in to take his place. Others in the Arnett family also worked in responsible positions for the ranch, including Wylie Daniel, nicknamed Tom; Albert Henry, nicknamed George; and Collom Bascom, nicknamed Bass.

After Ellwood and his wife died in 1910, the ranch was ripe for change. The Ellwoods' estate was settled with their two daughters receiving cash and the Texas ranch ending up largely the concern of W. L., who had been his father's confidante and partner in the operation from the beginning. The younger son, Perry, inherited ranches and other properties in Illinois, where he assumed control of the family banking interests.

In 1911 the business office of the Texas ranches was moved to Lubbock to be more centrally located to the ranch's two major divisions. This practice proved common in operating large ranches, though more often than not, the move was made to be closer to a central market. The office was initially located in Citizens National Bank in downtown Lubbock, but moved in 1929 to a building the ranch had constructed nearby at 1200 Broadway. The move to Lubbock stimulated interest in the city, and the ranch supported bringing Texas Technological College, now Texas Tech University, to the city. It was also instrumental in establishing the Ranching Heritage Center, an outdoor museum of ranch structures from across Texas ranch country.

New leadership included Otto Jones, son of W. D. "Black Bill" Jones, foreman of the H Triangle Bar Ranch near Maryneal and a longtime friend of Dick Arnett, who had hired Jones five years earlier as a cowboy. Jones would work at the ranch for some sixty years. Another important contributor was William F. Eisenberg, originally from De Kalb, the Ellwoods' Illinois hometown. He played a very important role in running the ranch from 1910 until the early 1970s. He died in 1972.

One of the battles Jones had to fight was against tick fever, a long-standing problem for the Texas cattle business. Regular dipping of cattle was the only solution, but the size of the Renderbrook's range and the large number of cattle made effective dipping difficult. All of the cattle had to be dipped within twenty-one days to break the tick's life cycle, and even missing a few head of cattle meant the ticks would escape eradication. The problem was finally solved in the disastrous drought of 1916–1918, when the ranch had to drastically reduce its herds. Entire pastures lay idle during the period, and deprived of the host animals for renewed cycles, the ticks died. In pastures where cattle still ranged, the numbers were small enough that the dipping was successful. Thus the apparent disaster ultimately proved a blessing of Mother Nature.

At the end of the drought, the ranch made a significant change from Shorthorn cattle to Herefords. It was a pattern mirrored in other areas of the West as well, because after the drought, ranchers noticed that the Herefords had had a higher survival rate than the Shorthorns. The Shorthorn breed has continued to decline to the point that it has all but disappeared in Texas.

Ranching on the scale of the Spade required a large number of good horses, and the remuda of the early 1900s included eighty horses at the south Spade, almost sixty at the north Spade, and at least five at the 49 Pasture. Supplemented by draft horses and mules to pull wagons, the herd swelled to over 150.

Although the Ellwoods had a long history in horse breeding, their experience had been with draft animals. The Ellwoods decided to buy horses for the ranch, and they bought them by the carload, usually sixteen to eighteen per car. The horses ranged from three to six years old and were supposedly broken but certainly not well trained. A number of outlaws turned up in horses bought in this manner, but the men were able to use most of them. Later the ranch tried horses with Arabian blood, a breed known for its untiring ability to run but not blessed with the innate cow sense necessary to excel in roping and cutting. Later some Thoroughbreds were used in crossbreeding, but for the last twenty years the ranch has had a band of twenty-five Quarter Horse mares and stallions from the Peppy and Doc Bar lines, well-known Quarter Horse sires. Today the eight cowboys on the ranch keep a string of four to six horses each, depending on the availability of mature, experienced horses and how many young horses need training.

The range was not fenced into manageable pastures at first, and more than one roundup per pasture was necessary to brand the calves each spring. "Gathers," usually a riding net of cowboys, covered ten to fifteen sections. The number of animals rounded up in such a sweep could be worked and returned to pasture that same day. During these roundups, the men ate their meals at the chuck wagon and slept on the ground in their "soogans," or bedrolls. Spring roundup lasted about a month. The summer work focused on doctoring screwworms, a problem that decreased in the late 1930s when the ranch started branding in cooler weather, when the threat of worm infestations was less.

By 1914 rail lines had reached the High Plains and made the north Spade property desirable for settlement. Always the astute businessman, W. L. Ellwood saw change coming and donated the right-of-way to the railroad rather than try to block its progress. Because ownership of the Spade lands was protected by deeds, the threat of losing access to the land because of settlers was not a problem. Those who had only some deeded property and ran cattle mainly on open range had to sell out and move.

Settlers came by the thousands, and plows turned under the sod as the High Plains were transformed for farming. The influx of people meant schools and roads had to be built, and services provided. These things had to be paid for with tax money, creating a burden for the Spade. W. L. eventually agreed to sell ranch lands at $35 per acre, considerably more than the original cost.

Ellwood Farms replaced the North Spade Division, and by 1926 35,000 acres had become farmland, and the cattle were gone. Within a year, W. L. had sold all of the northern part of the Spade and that part of the southern ranch suitable for farming.

RENDERBROOK SPADE RANCH

He sold the 49 Pasture as well. Development of farms continued until the Great Depression and the Dust Bowl, which slowed things down for a time. But that era passed, and as World War II loomed and the economy improved, farming on the High Plains moved into high gear. The heyday of ranching on the High Plains was over.

The South Spade, now called Renderbrook Spade Ranch, remained a ranch and has long been known for its quality cattle. The ranch also began running sheep in 1935, due in part to Perry Ellwood's insistence after the death of W. L. At first there were about ten thousand head. This change reflected a diversification in operation because the ranch then had three sources of income: lambs, wool, and cattle. The ranch proved that not only could sheep and cattle use the same range, but that sheep actually improved the range for cattle. Otto Jones noted that the sheep grazed broom weeds and other such plants, allowing better growth of grasses for cattle grazing. Sheep were phased out during the drought of the 1950s but were brought back in the 1960s. The ranch has about 2,200 sheep in good years, but runs fewer in dry times.

Another asset that has been valuable to the ranch is the development of oil leases. Such income has secured the position of the Renderbrook Spade as a continuing livestock producer and enables the ranch to continue its conservation practices. The ranch has long been a leader in brush control measures, particularly in chaining brush down, spraying it, and employing controlled burns.

The current generation of management has proved strong. Frank Chapell, an Ellwood descendant who ran the ranch, saw the need for sophisticated management and secured the services of W. J. "Dub" Waldrip, a new type of manager. Most of the previous managers had been trained in either business or life on the range, but Waldrip had a Ph.D. in ranch ecology from Texas A&M University. When J. E. "Shorty" Northcutt, Renderbrook manager for many years, retired, his son, Bob, took his place and currently runs the day-to-day operation. Under Waldrip's direction, the ranch began crossbreeding to improve its cattle and eventually developed a herd bred from Herefords, Black Angus, Brown Swiss, and Simmental. The Renderbrook Spade also keeps purebred herds of Hereford and Black Angus cattle for breeding, and some of these are shipped to other divisions of the Spade operation: the Borden Spade in Borden County; the Wagon Creek Spade between Throckmorton and Seymour; and the Chapel Spade near Tucumcari, New Mexico.

Eight cowboys ride the Renderbrook range. These men still take their meals at the cookhouse on days that they are doing roundups or other kinds of work as a group. All of them are married and live with their families in houses furnished by the ranch.

Renderbrook Spade Ranch stands today a viable cattle and sheep operation. Rendlebrock's spring still flows, and the "White House" erected by the Snyder brothers stands as part of a house at what is still the headquarters of this division. Because of management with good vision and ability, and the support of the owners, the Bassham family, who are descendants of Isaac Ellwood, the ranch continues to be known for family ranching values, quality livestock, and good cowboys.

Trans-Pecos Texas

LYING BETWEEN TWO MIGHTY RIVERS—THE PECOS AND THE Rio Grande—is a desert and mountain region that has been important to ranching for many decades. The Spanish explorers dubbed it *el despoblado*, the uninhabited place. Travelers going west on Interstate 10 or 20, which join west of the town of Pecos, find themselves in a vast openness that does, indeed, appear uninhabitable. Yet tucked away near sources of precious water—wells and springs—ranches have existed for more than a hundred years. The land next to the rivers, especially the Rio Grande, has been ranched even longer by Spaniards and Mexicans whose land grants always included access to sources of water.

This region is so remote that even shopping for groceries is a challenge. Often miles of gravel road separate a ranch family from a paved road that leads to a small, distant town such as Alpine, Crane, Van Horn, Marfa, or Kent. Distance makes communication difficult but even more important, and two-way radios and mobile phones have proved essential.

Early housing was constructed with adobe brick made from the soil. Some builders even used sotol, the tall, slender stalks of the cactuslike plant by that name. These stalks were set upright on each side of a board and the wall cavity filled with dirt to form a well-insulated barrier against the heat.

The land is harsh and unforgiving of abuse, and ranchers are careful in handling their pastures. So little grass grows in the desert that controlled burns are impossible in many areas. More significant, however, is that if the vegetation is destroyed, it could be years before ground cover reestablishes itself.

Plant life in the region is varied. Grasses include tobosa, grama (silver, feathered, and cane), blue, bluestem, tridens, plains bristle, wild rye, Texas fescue, Texas winter, and green sprangle-top. Creosote brush, greasewood, yucca, lechuguilla, catclaw, ocotillo, cholla, and prickly pear also thrive here. One of the most dangerous plants is locoweed, which is toxic to grazing animals.

In the mountains sparse but rich grasses provide forage for livestock and wildlife. Ranchman Ted Gray says the eroding red granite in the Davis Mountains gives the forage a high mineral content that stimulates animal growth. Other mountains in the area are Mount Emory and Lost Mine Peak in the Big Bend, and farther north are the Guadalupes, the tallest in Texas, whose highest point, Guadalupe Peak, has an elevation of 8,751 feet.

Between these mountains are desertlike basins. Near Van Horn, Wild Horse Draw empties to the north into a flat basin where rainfall leaches minerals from the soil and then evaporates. The salt thus deposited here and in other low basins is essential to many animals. This precious mineral also was collected by early settlers. The water of the Pecos River is so laced with salt and alkali that locals joke about its potency.

Animal life includes mule deer, antelope, javelina, some whitetail deer and elk, and aoudad sheep, which were introduced from North Africa. Cougars traveling north out of Mexico are numerous, especially in the Davis Mountains. In some areas, the large number of mountain lions poses a threat to survival of the deer population, their main food source.

Mining in the region focuses on talc, but there are also mercury, silver, and even limited amounts of copper, lead, and zinc. The Permian Basin on the eastern edge of the Trans-Pecos region is one of the major oil-producing areas of the world and is the main reason the towns of Midland and Odessa exist.

Important historically, the region is famous for the Goodnight-Loving Trail used to drive herds from Northwest Texas to the route of the Butterfield Overland Mail, which stretched in the 1850s from St. Louis, Missouri, to San Francisco, California. Its route led to Horsehead Crossing on the Pecos and on to California. Charles Goodnight and Oliver Loving turned their herds north at Horsehead Crossing and went on into New Mexico and Colorado.

The starkness of this landscape challenges the usual notions of beauty in nature, but to longtime residents, the area offers a life they would trade for no other. The unique demands of ranching here may require the best effort of managers, but at the same time the area provides a dry, healthy climate in which to live and work without a lot of people around. It is the least populated part of the state. The least populated county, Loving, has only a hundred or so residents, all of whom have to haul their drinking water from somewhere outside the county. Only one farmer remains in the county. Ranching offers the only viable activity for much of the region.

In no other region does the old adage about the importance of ranching only in an area where the rancher understands the full implications seem more apt. A rancher can get in serious trouble here by not recognizing the region for what it is—a potential graveyard for a cowman's dreams.

Henderson Cove Ranch

THE HENDERSON COVE RANCH LIES NORTH OF ALPINE IN Jeff Davis and Brewster Counties along the southern edge of the Davis Mountains in far West Texas. Elevations range from 4,300 to 4,800 feet. This altitude and location have a strong influence on the kind of ranching operation that can be conducted here because of the often scant rainfall and rocky soil, but cool temperatures and rich grasses. The current owner is Clayton Williams, Jr., of Midland.

Prominent landmarks on the ranch include Polk Peak, Black Canyon (also known as Dark Canyon), Anderson Spring, Benson Spring, and several sites where early settlers attempted unsuccessfully to homestead in this stark, demanding area. These latter spots are identifiable because of the discarded domestic items and walls of crumbling adobe, the primary building material used by early settlers.

The Henderson Ranch was founded in the late 1890s by the wife and brothers of William Thaxton (W. T.) Henderson in what has to be one of the most interesting sagas of ranching along the Mexican border. Henderson acquired much of his ranching knowledge growing up in the San Saba area of Central Texas. His father, James Elias, a Tennessean, was a veteran of the Battle of San Jacinto and served in Capt. Benjamin Franklin Terry's Texas Rangers. When his father died in 1883 and the heirs divided the estate, Henderson sold his part of the land, took his money, and, with his brother James, drove eight hundred head of cattle to the state of Coahuila, Mexico, to ranch there.

Henderson bought additional cattle over the years and established a partnership with the Stilwell family, already located in the area. The Stilwells rose to prominence in the Big Bend area of Texas and remain so today, the best-known recent figure being Hallie Stilwell, a ranchwoman and writer. In 1889, Henderson married Alice Stilwell, the first of three wives. They had two sons, Willie Gilmore (b. 1902) and James (b. 1904).

The Hendersons' fortunes in Mexico prospered for a time, but trouble began with a neighbor named George Cheesman. The initial conflict involved the ownership of some cattle. Further conflict with Cheesman eventually resulted in Henderson going to prison for stealing horses and mules belonging to Cheesman. Although the details are unclear, it is likely that Henderson simply took the blame for the crime of taking the stock across the border from Mexico to Texas to save another family member in poor health from serving time in prison. Transporting stock across the border was common, but apparently Cheesman chose to make it an issue in this case and had the legal clout in Texas to do so.

During the two years of Henderson's incarceration, his wife and her brothers established what became the Henderson Ranch on the Texas side of the Rio Grande. Upon his release, Henderson took control of the ranch and eventually increased his holdings to nearly forty thousand acres.

His dealings in Mexico, however, had one more chapter to play out. His old nemesis, Cheesman, was a key player. Cheesman, who apparently had contacts in the Mexican military, managed to have soldiers run the Stilwells and the Hendersons out of Mexico. Just as he was leaving, Henderson managed to trade his property to Mexican ranchers for some cattle, which he and his hands drove to the border. Once there, the animals had to be inspected for ticks before crossing the international boundary. The cowboys spotted some ticks on the cattle and asked Henderson to let them rope those and remove the ticks before the inspector came to look at the herd, but Henderson did not think the ticks would keep the cattle out of Texas. He rode to Marathon to get an inspector, who, upon examining the herd, refused to allow it to cross the river. Henderson told his men to simply let the cattle go. Natural instincts took the cattle back to their home range, and Henderson's Mexican empire crumbled. Although he and his men slipped into Mexico occasionally and retrieved some of the cattle in later years, he never fully recovered this loss.

After his first wife died, Henderson married Belle Black in 1915. After her death in 1939, he married Kate Espy, daughter of a highly respected local ranching family and widower of a man named Finley. His son Willie became a successful cattleman and kept the family name respectable. The other son, James, was a different matter.

Ted Gray, a legendary cowman in the area and owner-operator of part of the Henderson properties in later years, recalls that James was an excellent pistol shot and ran afoul of the law on several occasions. After killing a man in California, he fled to Mexico, where he married. W. T. was able to keep him out of prison only by paying significant sums of money to authorities. James returned to Texas and married again, but apparently not happily: he ended up killing his wife and her father. This time his father could not stop the wheels of justice; for the murders James served time in the state penitentiary in Huntsville.

After Alice's death, W. T. leased his ranch to another famous outfit in the area, the Kokernot 06 Ranch, still a major operation in the Davis Mountains. He later split his estate between his sons and their families. Ted Gray leased parts of the ranch and operated it for himself. Over the years Gray bought parts of it, including what became known as the Headquarters Ranch, which Berry Beal now owns. Another part is now owned by Al Micallef.

Ted Gray ran the ranch in traditional fashion. Even in the rough terrain, he used a chuck wagon,

and he carefully organized the work in sequence as the crew of sixteen cowboys worked its way across the ranch. Although it was sometimes difficult to move the chuck wagon and a hoodlum, or bed wagon, it was the only way to manage the work. Gray felt it was much better than trying to use pack animals, as some did in mountainous terrain or heavy brush. He recalled many a time moving the wagons before daylight and seeing the iron-bound wheels strike fire as they bounced off the hard rocks.

Gray believes the land around Alpine is among the best ranching country available. The gradual decomposition of the red rock formations provides minerals that make the various native grasses especially nutritious, and he has proof. Once he split a large herd of heifers by putting half in the mountains and half on the flats. Three years later, when they had developed into mature adults, the cows raised in the mountains weighed about 150 pounds more than those raised on the flats, and the calves weighed 50 pounds more than those out of the smaller cows. A good cowman watching the profit and loss margin notices such details.

Gray developed his practices for branding calves into a fine art. The cowboys would separate the calves from the cows and put the calves in a small pen, about a hundred at a time. Then the dismounted men would catch each calf by hand and throw it down near the fire used to heat the branding irons, where a crew of men was waiting to work the calf. The action progressed in a clockwise direction around the pen, and in an hour the crew could work about a hundred calves—a good example of how organization can save time and effort. On the Kokernot ranches and on his leased ranches, Gray was able to work the stock in the spring in less than a month. Now he sees ranchers using other methods that take longer to work half as many cattle. He is well aware that the work costs money that must come out of the profits, and that efficiency is the way to save money.

Working calves in this area is different from working where the soil is sandy. Because of the rocky terrain, the usual practice of roping and dragging calves by the heels to the branding fire is not wise. Dragging a calf for even a few yards over this rocky soil will cut the hair off and damage the hide. If roping and dragging is used on large calves, the cowboy must rope the calf by the head so that the animal can remain on its feet while being taken to the fire for working.

One problem in this mountainous region is locoweed, a noxious plant that, if eaten by livestock, causes the animals to lose their senses and die. It is particularly virulent in rainy years. Gray says that he has never been troubled with the weed because he learned how to work around it. His strategy includes not turning the animals out on pasture after they have been deprived of food for a time, because when they are hungry, they will eat anything, including locoweed. If they are not overly hungry, they prove able to avoid the weed. Gray said that those who work their cattle often or use the animals for practice with their cutting horses or roping and starve the animals for fresh grasses run a high risk of loss from this plant. Gray said that all areas have noxious weeds and that cattle raised in the region and handled correctly will survive.

According to Gray, horses raised in the region, characterized by rocky soil with steep inclines, are best adjusted to working here. When he began running the ranch for the Kokernot years ago, the mare band ranged on fairly flat pasture land. He immediately moved the mares and colts to the roughest section of the ranch so that they would be accustomed to the dangers and demands of the threatening landscape. In this environment the horses develop smaller and harder hooves, and the colts become adapted to the land on which they will have to work later.

A good horseman learns to handle his mounts in the most productive way. Gray's practice in moving the remuda of about a hundred horses on the Henderson and Kokernot Ranches from location to location was to drive the animals. The preferred method was to have two men handle the drive and to force the horses to walk so that they could graze and drink ample water along the way. If the men pushed the horses, it was easier to injure one, and the animals could neither drink nor eat sufficiently.

Gray preferred Thoroughbred-Quarter Horse crosses. He believes that it is more difficult to get the work done on the smaller Quarter Horses, which can "trot in the shade all day" and not cover the large tracts of ground necessary to support cattle in this area.

This ranch lies in cowboy country, but along the border the influence of the vaquero is apparent. Gray himself came to prefer a lariat from sixty to seventy feet long to rope wild cattle in rough country. Cowboys usually use ropes of only thirty to thirty-five feet. Gray dallied his lariat rather than tying hard and fast in the cowboy tradition because dallying makes it possible to throw the rope away if it becomes entangled or if the man needs to let go of the animal to avoid injury to himself, the horse, or the cow. Dallying also allows the roper to maneuver his horse, take up slack, or trip the animal. A man who roped constantly and rode long hours each day, Gray became one of the most adept cattlemen in the state. He says much of what he learned came from famous cattlemen in the area, and one particularly valuable friend was Joe Espy. Gray succeeded in the cattle business, not in the oil business, as many ranchers in Texas have done.

In the 1970s Clayton Williams, Jr., a Midland oilman, purchased the ranch, one of several he owns. Rather than limiting himself to one piece of range, Williams has purchased several ranches and leased others to operate. He currently owns High Lonesome (which borders the Long X Ranch), Loma Vista (he leases part of this ranch formed from the old Marfa Air Base), West-Pyle Ranch, and the Sullivan Ranch in Wyoming, his largest and, as he says, his "coldest." He runs yearlings on that northern property. He also leases the Tom Good Ranch north of Big Spring, a ranch owned by his wife's family.

Williams has long been dedicated to the Brangus breed, a cross of Brahman and Black Angus. For many years he had a breeding herd of purebred Brangus at a Floresville, Texas, ranch. These cattle are "growthy"—that is, of good size and conformation—and solid beef producers that gain size at an early age. He is a major producer of breeding cattle, which he sells to outside markets. He has sold his steers the last several years through video marketing, a method which has worked well for him because of the long distance the ranch lies from a major market. Half of the heifers go into the feeder market and the other half into the breeding market. He artificially inseminates these heifers, which he carefully selects for fertility. He keeps for his breeding herd or for selling as seed stock only heifers that breed in a relatively short period of time. About a fifth of these do not breed fast enough and are moved to the feeder market. He does give his first-calf heifers on droughty ranges another chance to breed. All others are converted to feeders and end up in the slaughter market.

Most of the cowboys who work these cattle are of Mexican descent and are directed by Chapo Ramírez, a man Williams considers an excellent cowman and foreman. The language of the ranch is a combination of Spanish and English, a situation Williams is comfortable with because he grew up in that kind of linguistic environment near Fort Stockton.

Calving times vary on Williams' ranches, so working the calves is not always done in one well-organized sweep through the various properties. The men who do the work live on some of the ranches, but the use of pickups and trailers strongly influences the method of covering the work. The ranch no longer uses a helicopter because, Williams says, "It takes the fun out of the work." He still enjoys gathering cattle with men on horseback.

Methods of working the calves typically involve

Guadalupe "Chapo" Ramírez, ranch operations manager

separating the calves from the cows and wrestling the young animals down, which Ted Gray convinced Williams is the fastest. On the ranch near Big Spring, however, the men prefer to rope and drag the calves for the practice and entertainment. The sandy soil allows the men to heel the calves. For larger calves, especially, roping and dragging is the most effective method because they are difficult to work by just wrestling them down.

When working the cattle, Williams uses, along with his regular employees, some temporary day workers and takes out the chuck wagon. It is a social as well as a work time, and the family and friends enjoy this traditional ritual of the range.

Wintering the cattle is fairly simple on the Henderson Cove Ranch. In the high, dry altitude, the grass cures readily to provide bulk for the cattle, and the ranch hands feed only protein supplements during the winter. Often this supplement is in the form of protein blocks because the roughness of the terrain makes other methods impractical. The hands haul the blocks to remote settings in four-wheel-drive pickups on roads cut into the areas. The men normally feed no hay.

Williams also creep feeds his calves. This practice involves arranging feed sources in pens and controlling access to those pens through openings too narrow for cows to pass through. The supplemental feed stimulates the growth of the calves. This practice is discontinued when rains return in the spring and pastures again produce grass.

The horse is a valued tool on the Williams ranches, and the breeding program is an important part of the operation. Clayton's wife, Modesta, is the driving force behind their horse breeding program, which includes a large number of mares and various stallions. After the oil bust of the mid-1980s, Williams bought several mares from a friend at a good price but agrees that today he and Modesta probably have more horses than they need. "You get attached to a horse," he says, "and that makes it hard to get rid of one of them." But too many horses means too much feed and money to support them. He recalled one uncle saying, "I'd rather see a prairie fire coming than a herd of mares and colts."

Only geldings are ridden by the men on the ranch, though Williams admits that mares are often smarter. He knows that trying to use both geldings and mares at the same time brings trouble. Most cowmen know that if a gelding is around a mare, he forgets everything he ever knew about working cattle.

The Williams' horses do not compete in the show ring or the cutting arena but are used only to work cattle. Ranch hands use saddles of no particular origin, and many of them furnish their own. The ranch usually keeps a number of utility saddles for visitors.

Watering the stock is a major concern in this predominantly dry area. Williams uses rotation grazing practices in which large numbers of cattle are ranged on relatively small pastures for a short period of time and then removed for an extended period. Consequently, he must furnish ample water to the animals in a variety of locations. When he bought the ranch, windmills pumped many of the wells, but these have been replaced over the years by electric pumps on the best wells. Pipelines carry water to areas where it is needed. Troughs are made of either galvanized steel or concrete. Williams prefers concrete but was not always able to build them when needed. He feels that wild game can water better from the lower concrete troughs, and he has long been a game conservationist.

Williams was trained at Texas A&M University as a conservationist and remains dedicated to that practice, especially water conservation and use. He has done extensive work on his ranches in eliminating water-hungry parasitic brush as well as constructing dams that, instead of restricting the flow of water, redirect water to areas where the soil is compatible with growing grass to nourish his cattle. His practices include root plowing land where the soil is productive and eliminating the two principal parasitic growths, mesquite and catclaw. Prickly pear, which plagues many Texas ranches, is not a serious problem here because it does not thrive at higher elevations.

Wild game abounds in the area. Mule deer and javelina are common, but commercial hunting is allowed only on the West-Pyle Ranch, which has one of the highest mule deer populations in West Texas. Instead of commercial hunting, each fall Williams takes his children, his customers, and his employees hunting, and takes a chuck wagon along to make it a gala social occasion.

Clayton Williams, Jr., and Modesta run a quality ranching operation that is only part of the conglomerate of business interests which they manage from their corporate offices in Claydesta Towers in Midland. The ranch continues to use the one-third circle W brand developed by Clayton's father years ago. Clayton says, "Ranching is a joy, despite the poor rate of return on the investment. But the love of the land is so strong that it overpowers your economic better sense. If I didn't love it, I could have made a lot more money in the business world. But I love it. And I love what it takes to succeed—fighting brush, improving the cattle, trying to outthink the weather. . . . I feel fortunate to be able to own land and ranch. I am doing now those things which I wanted to do as a child and did not get the chance to do."

In the case of the Henderson Cove Ranch, he chose a historic spread to which to devote his time, energy, and money. It remains today a productive part of the colorful tradition of ranching in Texas.

Hudspeth River Ranch

WATERED BY THE CLEAR FLOW OF DEVIL'S RIVER, HUDSPETH River Ranch lies at the overlap of four geologic regions. Rio Grande alluvial soil deposits lie along the river. The western side of the ranch shows evidence of the Trans-Pecos influence of rolling-to-broken plains. Chihuahuan Desert characteristics, represented principally by growth of creosote brush, are found as well. Edwards Plateau geology of hilly, canyon-cut limestone characterizes the area east of the ranch.

Much of this ranch's topography is limestone hills, which in some cases rise sharply hundreds of feet above canyon floors. About twenty miles as the crow flies north of the Rio Grande and about as many miles east of the Pecos River, the ranch property lies in Val Verde County approximately fifty miles northwest of Del Rio. Texas Highway 163 runs through the ranch.

The site of Fort Hudson, an army post established before the Civil War and reopened after the conflict, stands just to the south. Soldiers there were charged with protecting travelers on the road between San Antonio and El Paso from attacks by Indians. Amistad Reservoir, created by damming the Rio Grande, stands southeast of the ranch and backs water many miles into the rocky channel of Devil's River, which also feeds the lake.

Founded by Robert W. Prosser in 1883, when he bought the first nine thousand acres, Hudspeth River Ranch has a major resource. Prosser was attracted to the property because of the reliable water source flowing from the headwaters of Devil's River. Prosser and his business partner at the time, Herbert Fitzgerald, traveled to Mexico, bought cattle, and drove them back to stock the ranch. In 1885, he was the first in the area to begin fencing his land, a difficult task because of the limestone found beneath the shallow soil on most of the ranch. Many of the post holes were chipped out of the rock with a pointed steel bar.

In 1888 Prosser drilled the area's first water well to a depth of eight hundred feet. In 1903 he brought the first powered shearing machine to the region and that year sheared 40,000 sheep. Prosser eventually accumulated 117,000 acres of land stocked with sheep, goats, and cattle. He later helped organize the Del Rio National Bank and served as a director of the Lockwood National Bank, afterward the Frost National Bank, in San Antonio. He was, indeed, a man of many talents who helped shape the history of the region.

Claude Hudspeth, another remarkable man, bought the ranch in 1915 and established his headquarters in Prosser's original house along a creek which pours out of a rocky hill just west of the main channel of Devil's River. Although he later claimed to be a native-born Texan, Hudspeth was born in Monticello, Arkansas, in 1877. He proved early in his life to be an achiever. Although his family emphasized educating the children, the Civil War and its aftermath had interrupted that pattern. Family tradition holds that he left his home in Medina, Texas, at age nine to begin his professional career. Later, while recuperating from a bout of tuberculosis at an isolated ranch near Sheffield, on the Pecos River, he reportedly found a copy of *Plutarch's Lives.* The book inspired him to educate himself, which he set about doing. The family believes that he worked for a time for his uncle at the newspaper in Bandera and later may have been a printer's devil (apprentice printer) for the *San Antonio Express.* At age fifteen he established his own newspaper, *The Ozona Kicker*, in Ozona, a small community that would become the seat of Crockett County.

Hudspeth bought his first ranch south of Ozona and soon thereafter was elected by his friends as the region's Justice of the Peace. He told no one that he was only twenty years old, actually too young to serve. In 1901, he was elected to the state legislature as a representative. In 1907, he was elected state senator in a district that ran from El Paso to Fredericksburg in Central Texas. He served four terms as president pro tem of the state senate and in 1911 was appointed to the Penitentiary Investigating Committee. He led a radical reform of the Texas prison system to "abolish the bat" (whipping the prisoners) and institute the prison farm system.

After moving to El Paso while serving in the legislature, Hudspeth studied law by reading in the office of Turney and Burgess. In 1909, he was admitted to the bar. In 1917, the state honored Hudspeth by naming a county in West Texas for him. He was elected to the U.S. House of Representatives in 1918 and served with distinction until he retired in 1932. While in Congress, he purchased the Altuda Ranch in Brewster County and drove 1,400 head of registered Hereford cattle there from his ranch along Devil's River, a distance of 255 miles; not a single animal was lost in the three-week drive across difficult terrain.

One of the Hudspeth River Ranch's principal attractions is the Devil's River and the massive pecan, live oak, sycamore, and mulberry trees shading its banks. Some of the oak trees are estimated to be six hundred years old. Running clear and cold year-round, the river is the lifeblood for this arid area. Originally called the San Pedro River by the Spanish-speaking settlers, its name was changed, legend has it, by Texas Ranger Capt. Jack Hayes. He is said to have commented that its rough terrain, frequent flooding, and rocky bottom did not deserve to be named after a man of God but was instead better named for Satan himself.

The ranch is currently run by Claudia Hudspeth Abbey Ball, Claude's granddaughter. Although born in El Paso, she grew up in Del Rio, San Angelo, and San Antonio, as well as Hartford, Connecticut, where her father worked for Aetna Life Insurance Company. He was a 1925 graduate of Princeton University. Her mother, Elizabeth, Claude's daughter, attended the University of Texas and Sweetbriar College. A strong advocate of liberal arts education, Claudia graduated from St. Mary's Hall, an Episcopal prep school for girls in San Antonio, and then from Vassar College with a degree in Spanish, a language in which she had been fluent since childhood. Her career has included working at the Institute of Texan Cultures in San Antonio, where she was the first volunteer. From 1976 to 1980 she managed the Texas Folklife Festival, first held in 1972. This unique festival highlights and recognizes the many cultures represented in the state's diverse population.

Claudia's husband, T. Armour Ball, who died in 1990, was for twenty-seven years a probate judge in Bexar County, in which San Antonio is located. The couple had two children: Alice Ball Strunk, who is married to cattleman Billy Bob Strunk of Weimar; and William Armour Ball, who died of lung cancer in 1985. The Strunks have three children.

Several structures make up the cluster of buildings at Hudspeth Ranch headquarters. The original house that Prosser built proved to be too close to the river, and in 1932 a flash flood washed it away. Flash floods are common in this region of shallow soil, sparse vegetation, and rocky hills, and Hudspeth erected a second house further away from the river. The house was flooded with four feet of water in 1954, but it survived to serve today as a house for visitors and hunters. In 1983 Mrs. Ball built a spacious house even farther from the river. This structure was designed to accommodate her husband's 6'9" frame. The beautiful house has parquet hardwood floors, and some of the appointments include stuffed birds, for Mrs. Ball is an avid hunter. Included in the collection are pheasants from various parts of the world as well as wild turkey and other fowl. Mrs. Ball travels extensively, shooting wild birds such as grouse, pheasant, and red-leg partridge, and she has more than once attended partridge shoots in South Africa.

Also at the headquarters is the home of longtime foreman Homero Torres, who has worked for the ranch for more than fifty years. There are also barns and numerous other buildings, including a saddle house, a salt house, a barn for storing feed, a two-story apartment for hunters and visitors, a wild game processing room, and a small guest house.

When Mrs. Ball assumed control of the ranch, she reorganized the operation and placed a major emphasis on sheep. She reduced the cattle herd from two hundred head to around fifty, and she eventually decided to stock only Beefmaster cattle, a breed she feels is well suited to the region. She also runs some steers for short periods during each year that rainfall has been sufficient to grow suitable forage.

Mrs. Ball's ranch management program includes the use of cell grazing, adopted from the practices of Allen Savory, a leader in this method of range management. She has been to Savory's Holistic Resource Management School in Albuquerque, New Mexico, and has also attended the school of Savory's former partner, Stan Parsons, who advocates a slightly different approach to cell or time-control grazing that places a greater emphasis on economics. Mrs. Ball also depends heavily on the expertise available from the Texas A&M University Agriculture Extension Service.

Mrs. Ball continues to raise horses for ranch work, and the mares trace back to a sorrel Thoroughbred stallion placed on the ranch by the government remount program in the days when Mr. Hudspeth ran the ranch. Mrs. Ball has bred the mares to several Quarter Horse stallions and raises only enough horses to retain a remuda of about eight horses, five of which are currently mares broken to ride plus three broodmares. Ranchers in this area must raise their own horses or buy them from other local breeders because animals brought in from less rocky and rough terrain are too easily injured in this demanding landscape. It is also necessary to use horseshoes with cleats, called in Spanish "con tacón" (with a heel), on the rear to help horses negotiate the steep, slick inclines.

Claude Hudspeth advertised on his ranch stationery that he had the best sheep and goats available. "If you have any better than mine I will buy them," he promised. That tradition is continued by his granddaughter. The sheep are high-quality Rambouillets, which she shears twice a year. The sheep are run on the uplands, as far as possible from the stands of horehound, a particularly troublesome weed whose burrs stick in the wool and lower the grade at marketing time. These plants plague the river bottom but are not as common on the uplands.

The ranch also runs goats: Angora goats for their hair, and Spanish goats for meat. The Angoras' mohair is also susceptible to the burrs of horehound and other pesky weeds and must be protected as much as possible. The meat goats have no marketable hair and are ranged in the river bottom along with cattle to help with brush control.

Shearing of sheep and goats takes place in a barn so old that no one knows its age. A cement floor, obviously poured after the barn was constructed, bears the date 1923. The structure is still sound except for some rotted rafters supporting the roof. To solve this problem, the hands have placed rubber tires on the tin roof to hold the sheets of metal in place.

Marketing of the stock is well organized. A buyer comes to the ranch to buy the lambs, which are weighed on scales at the ranch and then trucked away. A shearing crew comes to the ranch to clip the wool and mohair, bag it, and haul it to Del Rio. San Angelo has proved to be the best cattle market, but Junction is the best for goats. Trucks haul the cull goats and sheep to Del Rio, where they are sold into Mexico, a fairly recent change. The ranch now sees an average return of thirty dollars or more for each cull instead of the fifteen dollars previously paid by U.S. buyers. However, devaluation of the peso impacts this market by causing prices to go up and down unpredictably.

In recent years, new methods have surfaced. Mrs. Ball still sends the lambs to the feedlot but now retains ownership through slaughter and the sale of the dressed meat. So far the practice has been profitable, but conditions can change. Also, slaughter of lambs in Texas has been slow to catch on, so Mrs. Ball joined a group of investors who constructed a processing facility in San Angelo.

Watering of livestock on parts of the ranch removed from the river requires that well water be piped into cement storage tanks and then into shallow cement troughs strategically placed across the ranch. Net wire pens surrounding the troughs facilitate trapping the animals in order to work them. Windmills pump some of the wells, but recently solar-powered pumps have proved quite satisfactory. In an area with a predominance of sunny days and often little wind, solar-powered pumps are more promising than windmills.

To the uninitiated, forage seems scarce in this country, and it is, in fact, far from lush. One must realize, however, that two kinds of vegetation are present, grasses of various types as well as brush that provides browse.

Winter feed is often cottonseed trucked in and shoveled out on the ground for the sheep and goats. This feed is a source of high protein relished by the stock. Sometimes, especially when the ewes are lambing, the ranch hands provide protein supplement blocks for the animals to lick at will. The "feeding frenzy" prompted by putting out cottonseed often causes ewes to abandon their young, and the lambs can become separated from the mothers and die. The use of the blocks eliminates this problem.

The region's many predators are especially threat-

TOP: *Claudia Ball*

ening to the young sheep and goats. Numerous bald and golden eagles migrate to the area in the winter, and these large birds of prey are devastating to this industry. At one time, Mrs. Ball kept her Spanish goats on the hills. However, the "kids," which are birthed year-round, attracted the eagles and kept them past their normal departure date for northern areas, usually around mid-March. The eagles then stayed on to attack lambs and Angora kids born in late March and April. Now the Spanish goats graze in the winter in the river bottoms and are less conspicuous to the eagles because of the trees and brush along the water course, and the birds leave earlier and take less of a toll on the young sheep and goats.

Additional predators include bobcats and coyotes, which the ranch tries to discourage by running guard burros with the sheep and goats. One surprising predator is the raccoon, which kills young sheep and goats and devours the entire carcass, even the hair and bones. One year, raccoons accounted for a 20-percent loss in the lamb crop on one part of the ranch.

Commercial hunting of game on the ranch—especially deer, javelina, quail, and wild turkey—has proved profitable, with hunters coming from as far away as Florida and Georgia. These outsiders really enjoy the sharp contrast in landscape, and most of the hunters return year after year. Fishing rights are also leased to area ranchers who do not have access to the river. Some members of the fishing club have permanent campsites. Through careful management of its various natural resources, the ranch generates substantial revenue in addition to its income from livestock. Unlike ranches in many other areas, this one lacks mineral income, such as that from oil.

One of several female ranch managers in the state, Mrs. Ball is somewhat philosophical about her role. Although she says she is not a "women's libber," she nonetheless feels that women make excellent managers because they are often more task oriented than men and are able to remain focused on the business nature of the operation, not just the everyday routine of the work. Mrs. Ball makes the point that ranching these days is a serious business and must be conducted that way in order to survive. It needs to be run like any other for-profit business with attention to the bottom line. When she took over management of the ranch, it was in debt, largely because her father's love of cattle had led him to overstock the range. By reducing the size of the cattle herd, increasing the numbers of sheep and goats, and using rotational grazing (a practice which her grandfather had used before Allen Savory was born), she has allowed the range to recover and has worked the ranch out of debt.

Mrs. Ball has sought to support the various causes important to her career. Consequently, she has given time and resources to several organizations. She belongs to and is a director of the Texas Sheep and Goat Raiser's Association, the Texas Wildlife Association, and the Trans-Texas Heritage Association. She served on the Development Board for the Institute of Texan Cultures in San Antonio and is a director of the Del Rio National Bank, the first woman to serve in that capacity.

Hudspeth River Ranch is a carefully managed spread of 15,000 acres, only part of the almost fifty sections that Mrs. Ball owns. She was content to lease one of her other ranches to a neighbor, but she has assumed control of it as well. A vigorous woman who responds positively to challenges, she evidences the best of the modern ranching tradition in Texas. A world traveler, bird-hunting enthusiast, and sometime socialite, she is a friend of ranching and ranch people in Texas and beyond.

Long X Ranch

THE LONG X RANCH, WIDELY RECOGNIZED FOR ITS elongated X brand, has survived in the Davis Mountains of West Texas for more than a century. Founded by George and William David (W. D.) Reynolds in 1895, a large portion of the ranch continues to be operated by a direct descendent of W. D. Reynolds. Once encompassing more than 300,000 acres, the ranch surrounded the town of Kent and extended north for some distance as well as south of Interstate Highway 10 and the railroad into the Davis Mountains, the northern edge of which is visible from the interstate. The Davis Mountains have two famous geologic features: Sawtooth, a rugged formation jutting hundreds of feet into the air; and Rockpile, a huge jumble of large stones.

The Reynolds family came to Texas in 1847 from Alabama and settled first in the eastern part of the state in Shelby County. Soon, however, they felt the call of the open plains and moved to Galconda, now called Palo Pinto, between Fort Worth and Breckenridge. Later they moved to a ranch along Gonzales Creek near present-day Breckenridge and began a ranching dynasty. This generation included eight children: John Archibald (who died in childhood), George Thomas, William David, Susan Emily, Glenn, Benjamin Franklin II, Phineas Watkins, and Sallie Ann. Sallie married J. A. Matthews, a ranching pioneer in the Albany area, and later wrote a book, *Interwoven: A Pioneer Chronicle*, detailing the life of the family, part of which was spent on Lambshead Ranch. The ranch near Albany was run for decades by her son, Watkins

Reynolds Matthews, who died in 1997 at the age of 98. Matthews family members continue to operate the ranch.

The Reynolds family had several confrontations with Indians while living in Northwest Texas. Fortunately, George and W. D. were able to make a good showing in a fight. George served the Confederacy in Company E, Nineteenth Texas Calvary, under Col. Nat Buford; W. D. served in the Frontier Regiment, a military unit that helped protect frontier settlers from Indian attacks. After soldiers were withdrawn from the region to fight in the Civil War, the Indians believed that the whites had abandoned their efforts to conquer the country, so they stepped up attacks on the invaders.

George suffered an arrow wound in his lower abdomen during a running fight on horseback with Indians along the Clear Fork of the Brazos River. Although he was able to extract the shaft of the arrow, the steel point remained lodged in the muscles along his back. He recovered and returned to his normal routine, but the point remained in his body for years. Later, as the noted trail driver Shanghai Pierce stood by, a surgeon in St. Louis probed for the head but was unable to find it. When Pierce cautioned the surgeon to stop his cutting before he "cut the man to the hollow," the surgeon ceased his efforts and Reynolds sat up. The contraction of the muscles caused the arrowhead to become visible in the incision, and it was removed. The steel point, the pistol which Reynolds' friend, Si Hough, used to kill the Indian that shot the arrow, and the silver-mounted bridle taken from the Indian's horse are on display in the National Cowboy Hall of Fame in Oklahoma City.

The Reynolds brothers established several ranches in Texas, including the X Ranch at Round Mountain near the Clear Fork of the Brazos River in current Throckmorton County, the 9 R Ranch in Scurry County, and the 7 Triangle Ranch in Shackelford County. They also established ranches in Arizona, New Mexico, Colorado, Wyoming, and Dakota Territory and drove cattle literally all over the West after George's initial drive of a herd to New Mexico prior to the Civil War. For that drive he used portions of the route of the Butterfield Overland Mail, whose stagecoaches carried passengers and mail from St. Louis, Missouri, to San Francisco, California, before the Civil War ended this operation. The route passed through the area along the Clear Fork and on to Horsehead Crossing on the Pecos River. There Reynolds left the Butterfield route and turned north into New Mexico. Later Oliver Loving and Charles Goodnight, both noted cattlemen and trail drivers, used the same trail, and their names were used to designate it.

The Long X was established when George and W. D. bought the initial fifteen sections of land and a house from Tom Newman of El Paso. Over the years they purchased additional acreage, and the family added to its holdings considerably as W. D.'s sons came of age and filed to homestead adjacent land. The Long X also came to include 100,000 acres of former XIT Ranch land in the Texas Panhandle. The extensive holdings of the Reynolds Brothers Cattle Company were operated as a single unit with headquarters in Fort Worth. This city offered all the facilities the company needed and had the advantage of being close to the primary market, the Fort Worth stockyards.

The headquarters of the Long X continued to be located at the original site south of Kent. After World War II, the family added more structures, put the original house and some additions under one roof, and added a flagstone patio constructed of stone from the ranch. They also built a spring house that used evaporating water to cool milk, butter, and other comestibles. A generator, known to the family as the "the Delco," furnished electricity for the headquarters until rural electrification reached the ranch.

Social life during the period, limited by modern standards, fascinated those involved. It included

horseback riding for the children, picnics, hunting, and visits with neighbors such as the Means, Cowden, and Jones families. The local polling place was the Long X headquarters, so election day brought families from a wide area to the ranch, where many stayed for a meal and a visit.

In later years, some family members, particularly the women and school-age children, lived in Fort Worth and traveled to the ranch by train. It was convenient and exciting to catch the train in Fort Worth at 10 P.M. and arrive at Kent about 11 A.M. the next day. They had their meals in the dining car and slept in the Pullman. When the railroad discontinued passenger train service, the family had no option but to drive or fly to Midland, the nearest airport, and drive from there.

For working the cattle, the ranch early established practices still followed today. Cowboys still rope and drag calves to the fire for branding, castrating, and vaccinating. Glenn Leech, who worked on the ranch during the 1930s, remembers that following graduation from high school in 1931, he and Jimbo Reynolds, a nephew of George and W. D., loaded their gear—saddles, bridles, spurs, boots, bedrolls, and other cowboy accoutrements—into a Model T Ford coupe and drove to the Long X to work as cowboys. Leech remembers it as one of the greatest adventures of his life. Their first job was to help construct a strong corral to pen a herd of buffalo that roamed the ranch. They then joined the crew of cowboys living with the chuck wagon as it moved across the ranch for the late summer and early fall branding.

Preparation for roundup was predictable. Each cowboy was responsible for shoeing the ten horses in his string. Since there were ten cowboys on the crew, the ranch's remuda contained more than a hundred horses, counting the extras needed in case one of the mounts became injured in the rough terrain.

The hands worked seven days a week and depended completely on the cook and the wagon for their support. Supplies from the ranch headquarters were brought out by truck or car to resupply the wagon, which moved at least every other day to a new section of the ranch. The cowboys slept under the stars in traditional soogans, bedrolls made up of a large tarpaulin intricately folded to provide protection from the cold. On nights when rain threatened, the men retreated to small, peaked rectangular tents designed to accommodate two men.

Food consisted primarily of fresh beef. When the cook ran short, the crew would select a fat, mottled-face heifer calf, one less desirable for marketing, and slaughter her. The diet also included pinto beans, Dutch oven bread, and very often, Leach remembers, a dish the cook labeled "bear-sign." A concoction served more as a dessert, its principal ingredients were canned tomatoes and sugar.

Since there were no pens for the horses, the night wrangler kept the herd together within easy distance and by daylight brought them to the wagon. There the men joined their lariats to form a rope pen and enclose the herd. Leech recalls that the animals cooperated readily and that he never saw any of the horses try to bolt past the single rope making up the enclosure. Once the wrangler gathered the remuda, the men entered the rope pen one at a time to catch the horses they wanted to ride that day and then saddled up.

During the 1930s, the Long X ranged between three hundred and four hundred broodmares on the northern part of the ranch. Because the mare band grew quite large at one point, the ranch traded more than a hundred mares to a rancher in Mexico for the same number of geldings. Gathering the mare band was quite a chore, with much racing after the wild horses in the rocky country on the north side of the ranch. After the cowboys had gathered the mares and separated the ones to be sold, the vaqueros from the ranch in Mexico took charge and drove them across the Rio Grande to the home range.

Breaking horses for use in the remuda took place during the winter months and included only geldings selected for work on the ranch or destined to be sold. Jimbo Reynolds recalls hearing about José Guerrera, one the famous bronc riders on the Long X. He came to the ranch as a fifteen-year-old boy and bet his only possessions, a pair of gloves and fifteen dollars, that he could ride any of the horses on the ranch. The foreman picked a particularly bad bucking horse and let the boy try. The horse won the contest, but José's efforts made such a good impression that the foreman refused the gloves and money and gave the plucky lad a job breaking horses for the ranch. He remained in that job for the rest of his life and died on the ranch an old man.

W. D. had several children who helped operate the ranch: George Eaton, Ella, William David, Joseph Matthews, Merle, Watkins Wendell, John, and Nathan. George had no children but adopted Betty Cantela, his cousin, who inherited his portion of the properties. W. D.'s children, however, had only five children: Joseph fathered Mary Fru; Merle (Harding) had Susan and Bob; Watkins had Watkins, Jr., who died at age thirty-three; and John had Susan Hughes. The ranch was thus eventually divided into only four large portions.

Hard times during the 1930s affected the Reynolds brothers along with everyone else. In order to hold on to their properties, they sold off the Round Mountain Ranch, the 7 Triangle, the 9 R, and the dairy farm, as well as most of their properties out of state. The Long X was kept intact, and after conditions improved by the early 1940s, the family began to add to the facilities at the headquarters as well as at several camps located at strategic points around the ranch. The only part of the ranch still in the family is 150 sections on both sides of Interstate 10, including the town of Kent. The ranch is operated by Mary Joe Goetzke Reynolds, granddaughter of Joe Matthews Reynolds, and her daughter, Lucy Weber.

Rick Furlong is the ranch manager, and he and his son, Clay, train the horses. Rick's goal is to raise good cow horses that can work in any kind of country. The colts spend their first two years pastured in rocky terrain. They are gathered as two-year-olds, broken to saddle, and then turned out until they are three, when they are finished. Rick and Clay are lifelong, accomplished horsemen. They train horses using a combination of old and new methods. Because they now work here, the Long X was invited to participate in the ranch horse competition at the Texas and Southwest Cattle Raisers Association foundation party.

The operation today traces back to the early days of ranching and the West. The mare band is still intact, though smaller than in the past. The reason is simple: the ranch needs fewer horses to do the work. The band produces about twenty colts a year, and the ranch keeps two to four of the horse colts to be gelded and broken to become part of the remuda of some forty horses kept for the cowboys to ride. The others are sold. In the late 1980s and early 1990s, the stallion with the band was a gray Quarter Horse named Stranger, a fine animal in whose veins ran the blood of Doc Bar and Poco Bueno, both famous sires. The ranch sold Stranger in 1992 and replaced him with a Paint stallion, Skip-A-Star, Jr.

A small herd of buffalo, descended from those Glenn Leech remembers building the corral for in the 1930s, still roams the Long X. Some are sold for breeding stock, and sometimes the ranch slaughters some for food. The creatures are prone to stay where they are content but can roam at will. Occasionally a young bull will be driven off by a larger, older male, and the rogue may decide to look for new range. Barbed wire fences are no deterrent to a rogue bull, and its roaming often brings it into contact with people startled by the sudden appearance of one of these shaggy brutes.

The mainstay of the Long X Ranch has been, and continues to be, Hereford cattle, though crossbreeding with Red Brangus provides some crossbred vigor. The ranch has tried other breeds in a limited

way, but the Hereford and Red Brangus seem best suited to the area's weather and forage.

Shipping cattle used to require driving the herds to the railhead at Kent. The cattle from the outlying areas of the ranch would be gathered near the headquarters for sorting and were then driven to the railroad corrals for loading. Then the cows, which were often driven with the calves to Kent, had to be returned to pasture. This was a demanding task requiring long hours and lots of riding.

After World War II, however, these drives stopped, and the ranch began hauling the cattle to the railroad corrals. After the railroad pens were torn down, the cattle were trucked to Fort Worth for sale. Now a company videotapes the cattle and markets them via satellite through a television network that allows buyers to examine the stock without having to visit the ranch. On the day arranged by the buyer and seller for trucks to haul the cattle, cowboys gather the herds and sort the animals to be sold. Within a matter of a few hours, the cattle are on the way either to other pasture or to feedlots in the Midwest.

Hunting also provides income for the ranch, which is supplemented by neither oil nor mining resources. Game animals include mule deer, antelope, aoudad sheep, and quail. Predators in the region are coyotes, bobcats, and badgers. Hunters from the more populous areas of the state to the east find hunting here challenging, exciting, and different from that closer to their homes.

One unusual facet of the Long X story is that the ranch owns the town of Kent, which includes a store. Goods include windmill and water pump supplies, livestock feed, some tack, supplies for area horsemen, groceries, soft drinks, beer, and gasoline. Also available are quality steaks from Long X cattle.

The remote setting of the Long X makes it seem like a lonely, inconvenient place to outsiders. However, those who live in the region adapt to and appreciate the high altitude, relatively dry climate, and, for the most part, pleasant weather. The nearest shopping is in Alpine or Pecos, more than an hour's drive away, certainly not a convenient arrangement, but residents just have to plan ahead better than city dwellers.

The Long X continues to be one of the dominant ranches in the state. Its teams compete successfully in ranch rodeos and in 1994 won the Western Heritage Classic Ranch Rodeo in Abilene. The history of the founding of this and other ranches by Reynolds Brothers Cattle Company is one of the most impressive to be found in the chronicles of ranching in the American West.

101 Ranch

THE 101 RANCH LIES IN THE CHIHUAHUAN DESERT NESTLED in some foothills cut with canyons about fifteen miles south of Van Horn in far West Texas. But the story of the 101 has more to do with the brand than, as one would typically expect, with a stretch of rangeland and a particular family name.

The ranch's present location in this starkly beautiful region does not reflect its origins. The story of the 101 Ranch begins in the 1820s, when four brothers—Abner, Robert, Joseph, and Peter Kuykendall—came with Stephen F. Austin's original three hundred settlers in the first legal Anglo migration into Texas. Abner is believed to have brought the first milk cows into the state. Joseph soon returned to Alabama, and Peter disappeared from family records.

Robert had a distinguished career with the Texas militia, forerunner of the famous Texas Rangers, and he rose to the rank of captain. After he was wounded in service, he was rewarded with land on the Texas coast in southern Matagorda County, on Tres Palacios River. He soon established a thriving ranch there and used the bow and arrow symbol as a brand. In the long run, this mark proved troublesome for two reasons: it often blurred or became indistinct on the cattle, and the large burn sustained when the brand was applied with the heavy irons often attracted blowflies, whose eggs hatched into screwworms.

Robert's son, R. H. (nicknamed Gill after his mother's maiden name, Gilliland), ran the ranch for many years. He passed it on to his son,

Wylie, who married Susan Pierce, the sister of the famous cattleman and trail driver Shanghai Pierce, and thus aligned two well-known ranching families.

Wylie served the Confederate cause in Texas and Louisiana and then returned to his ranching activities after the war. He dedicated himself to rounding up unbranded stock and soon had his ranch functioning again. He was active in the trail driving that characterized much of Texas ranching business during this time. By 1886 he was breeding newly imported Hereford cattle, still using the bow and arrow brand. Ranchers discovered that the British breed did not fare well in the hot, humid climate but that the Brahman, originally from India, did very well in the coastal region. Wylie realized the breed's advantages and became one of the earliest cattlemen in Texas to breed Brahmans.

According to family legend, when Wylie's son Robert married, he was given the 101 brand as a wedding present from the Miller brothers of Oklahoma, but the operation of Robert's 101 Ranch apparently had no ties with the Millers' famous 101 Ranch and their even more famous Wild West Show. Because the same Kuykendall family continued to run the ranch and simply changed its mark, the 101 Ranch can be said to have its roots in the earliest days of Anglo settlement in Texas. Early photographs of the 101 Ranch show cattle of a cross between Brahman, Longhorn, and various other breeds, all wearing the 101.

As the end of the nineteenth century approached, conditions dictated a change of location for the brand. By 1901, Wylie was making arrangements to trade for 13,000 acres encompassing several small ranches in Hays County near the present town of Dripping Springs. The family moved there by 1903. Family records do not indicate whether the cattle on the Matagorda Ranch were sold with the old ranch property or driven to the new property. In Hays County the family raised cattle and horses.

One of the stories about the history of the Hays County ranch involves a log cabin built that had been built there in 1838. Legend has it that Indians had killed a family living there, except for the wife of the man who homesteaded the property. Although scalped by the raiders, she survived and thereafter wore a scarf. In 1958 the cabin was dismantled and reassembled in Wimberly, Texas.

Another story involves hidden Spanish gold on or near the property. Around 1920, a neighboring family named Heimer suddenly had the cash to buy 11,000 acres of the 101 property, which came to bear their name. All of the neighbors believed that the Heimers had found the buried treasure and that this Spanish gold was the source of their sudden wealth.

The impact of the Great Depression on the 101 was serious, but family members held the operation together. The tenacity, vision, and hard work of Dorothy "Dottie" Kuykendall Hoskins deserves special recognition. During the Depression, the ranch kept a band of about eighty broodmares and raised horses, many of which they sold for use by the U.S. Cavalry. Dottie and her sister, Marian, broke the horses to the saddle before the army bought them. The Kuykendall women have always been exceptional horsewomen and rode English style, even when working cattle. Since military riding requires the same sort of saddle, training horses for the cavalry did not demand much change on the part of these enterprising women. Family records indicate that the women received $150 for "first choice" horses and $100 for "seconds," certainly significant amounts of money during the period, and difficult to raise any other way. The ranch also kept sheep during this time, the first raised by the family.

The Kuykendalls continued to ranch in this setting until 1966, when part of the family moved to a 23,000-acre spread traded from Robert Canning in Culbertson County and took the 101 brand with them.

The setting in West Texas is far different from the original one in Matagorda County on the Coastal Plain. This is unusual country. Throughout this re-

gion, the dominant flow of water is to the north, and all of the water flowing through the large draw or level lowland east of the 101 flows into a series of playa lakes to the north. The water that drains into these flat basins and evaporates leaves behind salt, which was valued not only for direct consumption but was also a precious commodity in a culture that depended on cured meat, not refrigeration, for preservation. In the 1870s an armed conflict broke out over rights to the salt.

The current operation of the 101 is managed by Elisabeth Hoskins; her son, Curt; and his wife, Katy Garren Hoskins. Elisabeth is the daughter of an Austin native who went to Europe before World War I to study art and married a German named Englehorn. After World War II, Elisabeth decided to migrate to the United States. To do so, she had to secure employment. She arranged a job teaching English-style horseback riding to children on King Ranch in South Texas, work at which she proved quite adept. Her experience at King Ranch included traveling with the family to such places as Jackson Hole, Wyoming, and California. She later met and married Laurence Hoskins, Dottie's son, who, with Elisabeth, ran the ranch for several years. They had three children: Mary Ann, Laurence, Jr., and Curt, named after Elisabeth's brother. Mary Ann sold her part of the operation and raises registered Black Angus cattle in Central Texas near Goldthwaite; Laurence, Jr., owns half of the land but is not active in management of the ranch.

Curt grew up on the 101, and Katy spent part of her youth on the adjoining ranch to the south, where her father's family lived. Curt, a graduate of Texas Christian University with a degree in business administration, feels that he is managing an operation compatible with his background. Also an adept mechanic, he keeps the ranch machinery running. After graduation, he worked for a time with a company in Indianapolis and feels fortunate to be able to live on the ranch. Katy, a graduate of Texas A&M University with a degree in agricultural sciences, teaches in the middle school in Van Horn and draws on her ranching experience as well as her science background in her teaching, especially in her course on Trans-Pecos ecology. They have one child.

Several structures dot the ranch. Elisabeth's home lies nestled in hills near the entrance, and Curt and Katy's house sits on high ground overlooking Wild Horse Draw, beyond which they can see the distant mountains dominated by Boracho Peak. About a mile east of Curt and Katy's house is a complex of barns with a set of pens Curt constructed of pipe. Nearby is a stucco house which once served as the main dwelling of the headquarters. It is currently used as a shelter for visiting hunters and as the cookhouse when cattle are worked, typically in the spring and early fall.

This site was originally occupied by a wooden house that figures in an interesting story. Around the turn of the century, the family of W. D. Smith lived in the house. After Mr. Smith, a stern taskmaster, refused permission for one of his hands to go to a dance in Van Horn, the man slipped off and attended the dance anyway. He became very drunk, and when he returned to the ranch, he killed Smith and his wife, and hid the bodies, which were never found. He then set fire to the wooden house, which was consumed totally in the resulting blaze. The man returned to town and proceeded to drink and brag about killing the Smiths. He was later arrested, tried, convicted, and electrocuted for the crime. It is said that the ghosts of the Smiths haunt the site, particularly the one bedroom that sits on the same ground as the original house. Some local residents believe that the Smiths hid gold on the ranch, but most people who knew the frugal couple seriously doubt that claim. Nevertheless, among the local beliefs concerning the site is the one that a witch, in the form of an owl that can assume human form when confronted, guards the place to keep potential looters from searching for the treasure.

Part of the ranch's income derives from leasing the hunting rights to a group of ranchers who like to hunt this particular section of property. The land supports mule deer as well as some antelope. Efforts have been made to introduce elk into the region. Aoudad sheep also flourish on the ranch. Predators in the area include numerous coyotes, which precludes raising sheep or goats. Foxes and badgers also inhabit the area. Game birds, especially quail, abound, though these are blue quail, not the more familiar Texas variety of bobwhite.

In earlier years, area ranchers shipped their cattle from the railhead at Valentine, a far easier drive than that over the rough terrain back to the railroad siding in Van Horn. In fact, Valentine was the mercantile outlet where most of the ranchers bought their supplies. Today most of the ranchers market their cattle via videotape through the increasingly popular satellite marketing system. Then the buyer sends trucks to the ranch to pick up the pasture-fresh animals for shipment to the desired location, usually pasture, before finally being shipped to the feedlot.

The Hoskins run Black Angus, a breed that does well in this part of the state. It is necessary, however, to provide strong mineral supplements to ensure good health and productivity. Vitamin A, zinc, magnesium, copper, and other trace minerals are provided in the form of blocks for the animals to lick.

Although the family has attempted raising horses, they found that, perhaps because of nutritional deficiencies in the grass, the results were unsatisfactory. The family has in recent years simply purchased horses, usually Quarter Horses or Quarter Horse-Arabian crosses, to do their cattle work. The family relies often on the help of neighbors, who pool their efforts in order to run their operations economically by not hiring outside help, which is very difficult to find and retain in this region.

Watering stock on the ranch is a constant concern. More than thirty miles of buried pipes conduct water to the pastures. Nowhere on the ranch must an animal walk more than thirty minutes to find water, but Curt must constantly monitor the troughs to make sure that ample water is available. The water level in the wells that supply the water is also a constant concern.

Part of the challenge of living in this region is the limited access to what most people have come to consider essentials, not luxuries. Shopping for groceries in Van Horn differs noticeably from shopping in a larger town or a city such as Alpine, and especially from that available in Midland or Odessa, some 150 miles away. Television service is limited to one channel available from Midland unless the family buys a satellite dish. Communication of all sorts is limited. The area once had a crude wire-line telephone service, but that system deteriorated to the point of being unreliable. Now the Hoskins family relies on a cellular phone. Family members and ranch workers also communicate through two-way radio.

Ranching in this region requires tenacity and careful use of the land. Since rainfall is scant, abused land just does not have the opportunity to recover. Some of the precious moisture is conserved by removing creosote brush and parasitic plants by root plowing or the use of chemicals, but controlled burning is ineffective because creosote is so acidic that no grass to fuel a burn grows around the base of the brush. Burns are also risky because of the possibility that the fire might escape the bounds intended for controlled burn. The area has limited rainfall, and if the rains do not come in the usual patterns in July and August, the burned range is devoid of foliage to support the stock. Controlled burning is useful mainly on tobosa flats and alkali sacaton draws.

Typical forage in the region includes tobosa grass in the flats as well as several other grasses in the grama family, especially black grama. The creosote brush offers no grazing potential at all for stock or wildlife. Some varieties of cactus thrive in the area, particularly cholla, known for its prickly spines. Wildlife browse some varieties of cacti.

Because of the area's remoteness, the long distances between ranches, and the relatively poor communication, personal relationships tend to be more valuable here than where the population is denser. Young couples in their thirties, such as the Hoskinses, are few in number. But when area residents can gather for social events or, perhaps more common, when they help each other work cattle, they especially value their friendships and contacts. They also have occasional contact with young ranching families from around the state, whom they see at such meetings as the Texas and Southwestern Cattle Raisers Association. All things considered, this is a unique culture, different in several ways from that found in other ranching areas of the state because of the unusual demands of this unforgiving but starkly beautiful land.

Conclusion

RANCHING CONTINUES TO BE ONE OF THE MOST important aspects of Texas heritage. Perhaps the most significant recent recognition of ranching's role in the state can be found in an issue of *Texas Monthly* magazine, a publication usually devoted to depicting the suave and chic, and debunking the Texas myth. However, the August 1998 issue contains a series of major articles on ranching topics, including one on a battle for power in King Ranch management, a profile of Dolph Briscoe as the owner of the most ranch land in Texas, a feature on the famous Matador Ranch, and a breakdown of twenty of the top ranches in the state. A serious look at ranching by *Texas Monthly* says volumes about the continuing financial, historical, and psychological importance of ranching to Texas.

Ranching practices have changed radically since the days of the vaqueros, who later taught English-speaking settlers from the southern United States how to ranch on the open plains. Along with their techniques came terms still used in ranching today and which trace back to Spanish—*ranch* from *rancho*, *lasso* from *lazo*, *chaps* from *chaparreras*, and many more.

Some of the most obvious changes occurred after the 1886–1888 drought and other droughts, floods, and market crashes. Ranchers learned to fence the ranges to keep the cattle they wanted to care for in one place. At first, fencing was just on the perimeter of property, but later interior fences, called cross-fences, were built to make herds even more manageable. Now the use of small pastures, called cells, through which cattle are regularly rotated, is becoming common. In this way, the rancher can run more cattle on fewer acres and hope to turn a better profit.

Over the years ranchers have bred better livestock, both cattle and horses. Crossbreeding of cattle has proved productive because a crossbred calf weighs more at market time than a purebred one. From the Brown Ranch with its Hotlander crosses to the Spade Ranches with their four-breed cross, crossbreeding has become the mode of the day. In order to keep these crosses alive, however, some ranches must keep purebred herds for breeding stock. The O'Connor Ranches' herds of purebred Herefords are a good example.

Horses are even more carefully bred, with great attention given to which bloodlines will produce the best working horses for cowboys to ride, or performance horses to be raced or shown in the cutting arena or show ring. Still treasured are the bloodlines of San Pepe, Doc Bar, Poco Bueno, Joe Hancock, and other great horses prominent in the annals of the American Quarter Horse Association.

The use of equipment has altered ranching practices enormously. From the earliest use of calf cradles or working tables that hold the calves so that cowboys can work the animals, equipment has allowed ranchers to use fewer and fewer cowboys and thereby lower expenses. The most radical changes, however, were wrought by the development and use of the pickup and trailer for hauling livestock, which further reduced the number of hands necessary to do the work. The helicopter has allowed further reduction of personnel, but the rancher reaches a critical point when not enough people are at the pens to work the cattle when the helicopter pens them.

The clearing of brush by burning and chemical means has raised the carrying capacity of pastures, and the planting of native grasses and the growing of winter wheat have provided increased grazing as well. Range management remains a constant concern and expense for ranchers as they strive to do one of the basic tasks of ranching, providing the food necessary for the herds to thrive.

Economics has become the watchword for the modern rancher. Access to oil income has helped many a rancher survive poor management decisions, drought, failed markets, and other factors that often bankrupt a business of this sort. However, the collapse of the oil business in the late 1990s was a serious blow to many ranchers. The leasing of hunting rights to outsiders has provided much needed income, and finding other types of supplementary income continues to be a concern.

Most interesting, perhaps, are those things that have not changed. The emphasis in ranching remains on raising quality livestock and making a profit. Ranchers also take care of the land because the land is what will see the operation through the years to come.

One constant in this ever-changing world of ranching is the cowboy, and horses continue to be his priority. Frank Graham, a long-time South Texas ranch foreman, says, "If a man does not enjoy the horses—raising, breaking, training, and riding—he will leave the work and find a job in town." I have heard cowboys inquire about the fate of horses they rode on ranches where they worked years earlier. The men were as interested in the horses they recalled from the ranch as they were in the men with whom they worked.

Ranchers still believe in raising horses in the area where they will work. Ted Gray, a Trans-Pecos rancher, always made a point to put his mare band and colts in the roughest pasture so the colts would learn early how to cope with rocks and steep grades. Also the high, thin air of mountain altitude taxes horses not accustomed to it. A horse must also become familiar with everything it will see in the pastures so as not to be startled by it and panic.

Cattle are still the reason for ranching but are at the same time the bane of the cowboy's existence.

Seemingly stupid by nature, cattle can often be vexing and frustrating. If one can say that cowboys love horses, it is just as true to say that the cattle draw their ire to the same degree. Cattle, like horses, must adapt to regional factors. A cow brought from the rich grasses of West Texas to the Victoria area, for example, will not know at first how much of the less nutritious grass she must eat to retain her body weight. If the process is reversed—from east to west—the cow will gain weight because of the increased nutrition.

The most important equipment for working cattle has not changed. Ropes are now nylon rather than the earlier grass or manilla, but it is still a cowboy's most treasured tool. One Texas writer, John Erickson, devoted an entire book to ropes and roping methods, titled *The Catch Rope*. The saddle, the cowboy's leather throne, is still handmade and built to take the strong pull of the taut lariat around the neck or horns of a large bovine. The chaps protect the rider's legs and show personality as well in the length, cut, and decoration. The spurs, often handmade and ornamented with initials and a brand—or, less likely, the hearts, diamonds, and clubs of an earlier day—still jingle on the heels of handmade boots with tall heels and tall, brightly colored and stitched tops.

The most important factor of ranch life is still the attitude of the cowboy. His life has changed, but he fights to keep the traditions alive and working. George Peacock, manager of the Nail Ranch, says it's the only life he wants to lead, and he feels lucky to have lived it all his life. Terry Moberley, foreman of the neighboring Lambshead Ranch, says that he still enjoys working outdoors and that the life of a cowboy is a good one. Today's ranch hands may be better educated than their predecessors, but they are still just cowboys at heart. And no matter where in the American West they work, their routines are the same.

Imagine, if you will, getting up by 3 or 4 A.M. in the dark of early morning to join a bunch of cowboys at breakfast at the cookhouse or around a chuck wagon. The clink of spurs is part of the music of the morning. Amid the storytelling and joking is a camaraderie that is as old as humanity. Then you and the others grab your tall hats and head for the corral, where the wrangler has penned and fed the horses selected for work that day. These are not stable horses. These are Texas cowponies, mature Quarter Horses trained and ready for work. You catch your horse, bridle it, and, in the dim light of a single bulb from the saddle house, get your gear straight and your horse loaded in the gooseneck trailer behind the four-wheel-drive pickup.

After a drive over rough gravel roads, or just a pair of tracks through a pasture, you unload, check your girths, and swing into the saddle. After listening to the foreman's instructions about where you'll be working that day, you ride to high ground and take in the view just as the sun begins to lighten the eastern horizon. You stop and breathe and smell the clean air tinged with grass and whatever season is upon the land—freshness in spring, curing grass and leaves in the fall, a damp wind in winter. Benny Peacock, foreman of the Bob Green Ranches around Albany, said it best: "If you can see that and not feel good about life, then you are probably already dead."

The rest of the day goes fast, riding hard on the roundup, working cattle in a dusty pen, gulping down a quick lunch, and finishing up before dark. Tomorrow, you'll do it again.

Many revisionist historians have declared the cowboy dead and gone. I have to disagree. I have had the experience described above several times in the last decade and a half as I have visited ranches and worked alongside these men from the Brazos to the other side of the Pecos. The men and women who cling to this life are a rare breed. They tend to be shy at first, but when you get to know them, they are warm, open, generous people, the best I have ever had the pleasure of knowing. They love their life, and they are fiercely proud of it. They prize their independence and their honor.

What does the future hold for the ranching way of life? One can believe that as long as there are cows, it will take cowboys to see after them, but what the life of that cowboy will be like is uncertain. Clearly, much of the land that was formerly used for ranching is being converted to other uses, particularly for housing developments with such names as Mesquite Ranch Estates. New ranches, particularly large ones, are not likely to emerge. With the small return on an investment in ranching, it is virtually impossible for someone to borrow money to buy land, livestock, and the like and ever have a chance of paying it back.

Those entering ranching now are typically attracted to its romance but will have earned their money from other sources. The oil business has been a mainstay, but an unpredictable market makes it more likely that those buying ranches will have made their money in business, medicine, or law. The families who opt to do this must be willing to spend their money on ranching rather than elsewhere.

The breeding of animals has become more scientific and will no doubt continue to do so. Artificial insemination and embryo transplants will continue to be used, and perhaps cloning will become common. Other technologies will develop as well. Better genetic control of herds to provide desirable characteristics will become even more important than it is today.

Smaller, better-managed operations will be necessary for survival. The average cattle herd in Texas is around fifty animals, with the large ranches having many more, of course, but this figure suggests that many people are in ranching on a small scale. Better management of the land and improved methods of brush control and conservation will become more and more important as ranchers have less land to manage. More than likely, recreational use of ranch land will grow. Already city dwellers pay ranchers so that they can participate in traditional cowboy activities, and people are also willing to pay for the right to hunt and fish on ranch properties.

In a business such as ranching, affected by both internal and external factors, careful management is crucial. An element of good fortune—just plain luck—is also required to survive. Even oil money or supplementary income from hunting will not guarantee success. It takes grit, guts, and vision to ranch, and those who have faced the challenge have created a large part of the mystique that is Texas.

Glossary of Ranch Terms

THE FOLLOWING GLOSSARY INCLUDES TERMS USED TO DESCRIBE and communicate functions in ranching, which, despite the romance often associated with it, is a business. Like any other specialized business, it has its own vocabulary. There are variants for some of these terms in other parts of the United States, but these terms are in general use in Texas. As in all specialized vocabularies, some of the words given here—*prowl*, *tack*, *string*, *crease*, *set*, and *bit*, for example—have variant meanings in other areas of life. Terms appearing in italics in the definitions are defined in their alphabetical order.

The root language of much of cowboy and ranching terminology is Spanish, the language of the *vaqueros* (cowboys) and *rancheros* and *hacendados* (ranchers) whose techniques were picked up and adapted by later immigrants to Texas. The Mexicans had been there for decades raising cattle and herding them with skilled riders on horseback whose skills with the *lazzo,* or lasso, were extraordinary. Many Spanish terms found in South Texas, which constitute an extensive expansion of the ranching dialect, are not included here.

The rich cultural background of Texans in ranching owes much also to a southern influence with British roots that brought a somewhat different view of cattle ranching to Texas. From this melding of cultures came the various techniques and practices for ranching in Texas.

Angus—a British breed of cattle popular on Texas ranges. They are hornless and a solid color, either red or black.

Annual cycle—the various seasonal events and chores collectively required to take care of cattle during each year.

Auction ring—a complex of pens and buildings where stock is auctioned to the highest bidder, usually one day a week, known to stockmen as Sale Day.

Barbed wire—usually two strands of metal wire twisted together and including sharp metal barbs. It is the widely accepted material for fencing pastures and the subject of much lore in the West.

Batwing chaps—a style of chaps with full-cut legs.

Bedroll—composed of a tarp used as an outer cover for sleeping and for enclosing blankets, clothing, and personal effects.

Bed wagon—extra wagon used to haul bedrolls on large ranches. See *hoodlum wagon.*

Beefmaster—a breed of cattle developed in South Texas in the 1930s by crossbreeding Hereford, Shorthorn, and Brahman cattle.

Bit—a metal device for controlling a horse by putting pressure on its mouth. The side bars are attached by a bar that goes through the horse's mouth. A low or shallow port bit has a gentle or small curve in the center; a high port bit has a sharp, high curve. There are numerous variations of the basic bit.

Bosal—a band made of plaited leather or rope, often reinforced with wire, that fits over the horse's nose above the animal's muzzle. A rope or reins are attached to it to control the horse if the bosal is used as part of a *hackamore.* Often referred to as a nose band.

Brahman—the breed of cattle from India. Characterized by medium-length horns, large ears, and a prominent hump on their shoulders, they have proved particularly resistant to parasites and disease on the humid Coastal Plain, where European breeds do not thrive. Coloration runs from white to grayish white with some darkening on the head and shoulders. Unlike European breeds, Brahman cattle have sweat glands.

Branding iron—a device with a handle three feet long or longer with the emblem representing the ranch reversed on the end of the bar so that the burned mark on the hide can be read normally.

Brasada—a Spanish term to describe thick growth of various thorny brush found in South Texas.

Breast collar or harness—a strap of leather, or, less often, of mohair, attached to the front cinch rings of the saddle and encircling the chest of the horse. The breast collar is most useful to keep the saddle from slipping to the rear when the roper is pulling an animal by a lariat tied to or dallied around the saddle horn.

Breeding stock—livestock used to produce offspring.

Brindle—a pattern of coloration. It is usually in the form of vertical dark stripes on crossbred cattle.

Bronc—An unbroken horse, or one that is not trained to be ridden or to work cattle.

Brood stock—See *Breeding stock.*

Brown Swiss—a European breed of cattle bred mostly for producing milk. There has been some limited use of this breed in Texas to try to develop cows that produce offspring with desirable beef characteristics that develop better because of the extra milk.

Browse—nutrients available to cattle from plants other than grass, usually leaves and forbs of brush or trees.

Brush jacket—a waist length coat of heavy canvas worn to protect the rider from thorns on the brush.

Brush popper—a cowboy wise to riding in the heavy covering of brush, especially in South

Texas. A good brush popper will come out of the brush with the animal he went in to bring out, though he may be carrying a small load of wood around his saddle horn.

Buckaroo—the version of the cowboy found in the northwest United States, especially Nevada, Idaho, Oregon, Washington, and California.

Calf cradle—a term sometimes used to identify a *working table.*

Camp—usually a barn, corrals, horse pasture, facility for storing feed, and a house where a cowboy and his family live. The man is usually responsible for the pastures that surround the residence.

Cantle—the rear part of the seat of the saddle. It may be short with the leather "rolled" back for a *rolled cantle*, or it may rise up several inches to form a *straight cantle.*

Cattleguard—a barrier located where roadways pass through fences. Constructed usually of steel pipe, the floor of parallel pipes with space between covers a shallow pit beneath. Livestock will rarely attempt to cross this kind of barrier, over which vehicles can pass without having to open gates.

Cell grazing—a range management concept that includes small pastures in which livestock are grazed for a relatively short period of time and then moved to other pasture. This high-intensity grazing for a short period encourages the animals to eat both desirable and undesirable growth and spread manure on the land, which is then allowed a period of rest before being grazed again.

Chap guards—a part of the shank on some spurs designed to keep the bottom of the chaps from touching the rowel on the spur.

Chaps—coverings for a rider's legs. Made of a soft but durable leather, chaps resemble pants with the seat cut out. See *shotgun chaps*, *batwing chaps*, and *chinks*. Also called leggings.

Chin strap—a small strap of leather, chain, or other material attached to the bit and fitted below the horse's bottom lip. It helps control the horse.

Chinks—chaps that reach just below the knee.

Chute—1. an area two to three feet wide between two strong fences. This area may be several yards long. The chute is used to force cattle to form a single line in order to be worked or sorted. The various working tables can be attached to one end of the chute, and the cattle will be driven into the chute from the other end. 2. The narrow pen from which bucking stock are released for that eight-second ride in a rodeo arena.

Cinch—(n) the strap, often of braided mohair or horsehair, that passes beneath the horse's stomach to hold the saddle in place. See also *girth*; (v) to pull the girth tighter to hold the saddle on the horse more securely, also called "cinching up."

Clydesdale—a breed of draft horse from Scotland.

Concho—silver medallion used for decorating chaps, saddles, bridles, and other gear.

Controlled burns—a conservation practice of selectively using fire to eliminate or retard the growth of parasitic plants.

Corral—one of the names for a pen or set of pens for holding or working stock. From Spanish *corrales.*

Cow—(n) the mature female of cattle; the instinctive and trained response of horses to the movements of cattle while working in a pasture or pen.

Cow-calf operation—the practice of stocking a ranch with mixed sexes in order to raise calves. See *steer operation.*

Cow camp—a temporary or semipermanent setup for the purpose of working cattle. It may be only the chuck wagon and the bedrolls with or without various kinds of tents. Cow camps are still used when the range to be worked is a great distance from the headquarters or the crew desires to emulate an earlier style of working cattle.

Cow horse—a horse with cow sense, trained to work cattle.

Cow sense—innate tendencies, found especially in Quarter Horses and some breeds of dogs, to work cattle with little or no training.

Coyote—an omnivore prevalent on the ranges of West Texas and across the West generally. Its melodious howls at sundown are the music of the range. The animals prey on small game and young deer as well as mice and rats and consume carcasses of dead animals. They have been known to kill small calves and even disabled cows unable to fight them off. Sheep and goats are especially susceptible to these predators.

Crease—the pattern impressed into the crown or top of a hat.

Crew—the assembly of ranch hands whose job is to work livestock. It may include temporary day workers and others from nearby ranches who are "neighboring."

Crossbreeding—the practice of breeding individuals of different breeds to produce offspring that have a combination of their traits. Crossbreeds are often more vigorous in growth than purebred offspring.

Cross-fencing—the practice of fencing large pastures into smaller ones in order to better control livestock.

Cull—to identify and remove inferior, old, and unproductive animals from the herd.

Cut—(v) 1. To castrate male animals. 2. To separate individual animals from a herd; (n) the cattle removed and held separately from a larger herd.

Cutting arena—a large pen, preferably with sandy soil and no rocks, in which to ride and train horses to separate cattle.

Cutting horse—the "ballet dancers" of the ranch horse world, these intelligent, agile animals are adept at separating and keeping animals separated from a herd. These are usually small-boned horses with small frames, not the type preferred by cowboys for the heavy pasture work of roundups and roping and dragging calves at branding time.

Dally—from the Spanish *de la vuelta*, meaning to wrap the lariat around the saddle horn when roping stock. The alternative is to tie the rope to the horn. See *hard-and-fast*.

Day work—doing ranch work by the day rather than being hired by the ranch on a full-time basis.

Dehorn—to remove the horns from cattle, usually with a dehorning saw on older animals or a variety of "spoons" or other devices on calves. Removal keeps the animals from hooking each other when in pens or hauling trucks.

Double-rigged saddle—a Texas-style saddle with two girths—one attached to the saddle below the fork or swell and the other below the cantle. This style is still preferred by Texas cowboys. Also called a "rimfire" saddle.

Draft horse—any of the large breeds used primarily for pulling wagons and plows.

Drought—an extended period in which rainfall is significantly below normal.

Feed run—the route followed in distributing feed to cattle during the winter.

Feedlot—a business operation in which cattle are fattened for slaughter. These are usually in grain-producing areas. It is more economical to haul the cattle to the grain than to haul grain to the cattle.

Flank girth or cinch—a heavy leather strap attached to the rear cinch rings on a double-rigged saddle. It passes beneath the horse's stomach and keeps the rear of the saddle from being pulled up by a taut lariat holding an animal located in front of the horse.

Flanker—a cowboy whose job it is to throw the calves to the ground for working when the crew is roping and dragging.

Flanking calves—the technique a cowboy uses to throw calves to the ground in order to work them. The man grasps the rope around the calf's neck with the left hand and the loose skin in front of the right hind leg with his right hand, and lifts the calf off the ground in order to throw the animal to the ground.

Foot rot—a condition in livestock in which the feet develop a condition of decay. The condition is common in areas of persistent dampness in East and Southeast Texas.

Forage—an inclusive term for all plant life available for cattle to graze.

Foreman—the man in charge of the cowboys on a ranch. He gets his orders from the owner or manager.

Fork—that part of the saddle that forms the front. These are usually slick (with no swell) or swelled (with a noticeable swell).

Gooseneck trailer—a device pulled by a truck and characterized by a hitch which attaches to the bed of the truck over the rear axle and arches over the tailgate. This arching tongue resembles the neck of a goose—hence the name.

Grazing—the act of eating forage by livestock. Also the plant life available to the livestock.

Green-broke—said of a horse that has been taught to accept the saddle but not trained for cattle work. These young horses may still pitch, or buck, unexpectedly.

Hackamore—headgear used for training young horses, made of a bosal and a leather strap for a *headstall*. Some riders prefer the hackamore for older horses as well.

Hair goat—any of the breeds of goats, usually Angora, raised principally for its valuable hair.

Hair pad—a protective covering usually placed between the horse's back and the saddle blanket to prevent sores caused by the rubbing of the saddle during riding. The hair pad is constructed of soft animal hair sewn to a sturdy cloth back.

Hand—a unit of measurement equal to four inches, used especially for the height of horses.

Hard-and-fast—the cowboy style of attaching the lariat to the saddle horn by tying the rope either by half-hitching it or using a slipknot or device made especially for the purpose (a *honda*). The alternative style is the dally method. See *dally*.

Head and heel—the method of roping an animal first by the head by one roper and then by the hind leg or legs (called the heels) by a second roper, thereby causing it to fall to the ground.

Headquarters—the residence of the boss and the foreman as well as the central offices, main horse herd, and other centrally significant operations. "Branch offices" for the ranch are usually called "camps."

Headstall—the part of the bridle that attaches to the bit and goes over the horse's head.

Heifer—a young female bovine, especially one that has not had a calf.

Hereford—a British breed of cattle that has proved very adaptable to many regions of Texas. They are characterized by red color with white faces. The breed may be horned or polled (hornless; also called *muley*).

Hobbles—leather, rope, or rawhide straps used to secure the front legs of a horse to keep it from wandering.

Honda—a metal ring used to form a slipknot to fasten the lariat to the saddle horn; or the small loop of rope, metal, or rawhide on the end of a lariat rope through which the rest of the rope is passed in order to make a loop for throwing.

Hoodlum wagon—the wagon used to haul bedrolls, tents, and other gear on roundups if the chuck wagon cannot hold all of the baggage. Also called bed wagon.

Hoolihan—a type of loop used by cowboys, most often to rope a horse. It is thrown backward from the regular loop and has the advantage of reaching its target at a ninety-degree angle with the ground; hence the loop is upright, and the opening is at its largest when going over the horse's head.

Horned stock—cattle with horns. Breeds normally with horns include Herefords, Brahman, and Longhorn. See *muley*.

Hot roll—a variant name for bedroll.

Hybrid vigor—the tendency of crossbred stock to grow rapidly.

Jacales—small brush huts commonly used by early settlers in South Texas.

Lariat—rope; often called by one or the other term (sometimes both), the lariat is one of the basic tools of the cowboy's trade and is usually thirty or so feet in length and made of ⅜-inch nylon.

Lasso—a term derived from the Spanish *lazo*, for the cowboy's lariat.

Latigo—the leather (less often nylon) strap used to tighten the front girth of the saddle to hold it securely on the horse's back.

Longhorns—the cattle originally brought to North America by the Spanish explorers as mobile food supplies. These cattle with unusually long horns served as the basis for early cattle ranching in Texas. Now a registered breed, they are often bred to first-calf heifers.

Mare band—the herd of broodmares for raising colts on a ranch. The Spanish equivalent is *manada*.

Mesquite—a parasitic thorny perennial plant, ranging from bush to tree size, with a huge appetite for moisture. Because of its nature, it is all but impossible to eradicate unless the base is dug up. The plant produces a bean palatable to livestock and, for Indians, the source of a kind of beer as well as a flour for bread.

Motley-faced—a pattern of coloration on a bovine's face, usually dark spots on a white face.

Mountain oysters—the name for calf testicles cleaned and cooked, usually dipped in corn meal and fried. These are considered a range delicacy. Also called calf fries and prairie oysters.

Muley—cattle without horns. Breeds that are naturally polled or muley include Angus and Polled Herefords.

Mustang—feral horses running free in the wild. The original mustangs in the United States were descended from the first horses brought to this continent by Spanish explorers and settlers. Although they served as the basis for transportation in early days, mustang blood is not considered desirable.

Navajo—a brightly colored wool saddle blanket. It may or may not be made by Navajo Indians, but the term traces from a time when these typically were made by Navajos.

Neighboring—a time-honored practice of the cowboys on neighboring ranches helping each other work cattle.

Net wire—steel wire woven to resemble a net. It is widely used in conjunction with raising sheep and goats. If properly constructed, a fence constructed of net wire can keep predators away from the sheep.

Night horse—a horse kept in a corral overnight so that the *wrangler* can ride the animal to gather the *remuda* in the early morning. It is often the only horse kept in a pen overnight.

Outfit—a term indicating anything from the cowboy's clothing to the ranch itself.

Outlaw cattle and horses—animals that refuse to be gathered and worked with the other stock. The best treatment for an outlaw is to sell it.

Ox-bows—a style of stirrup of iron, wood, or fiberglass characterized by a round rather than flat bottom. It allows the rider to insert his foot into the stirrup up to the heel of the boot. It is known as a "widow maker," because it is difficult for the rider to remove the foot if bucked off.

Parasites—Any of a number of organisms that draw life from livestock, both inside, such as various worms, and outside, such as numerous blood-sucking flying insects.

Parasitic brush—the forms of plant life that suck water from the ground and spread perniciously over the landscape. Varieties across the state are prickly pear and mesquite, but in South Texas include black brush and huisache. In dry areas and times of drought, these plants may be the only form of browse available to livestock and are, hence, not always undesirable.

Pasture-wise—said of cattle that have learned to use obstacles such as brush, draws, streams, and the like to elude cowboys during roundups.

Percheron—a breed of draft horse from France.

Pigging string—a short rope used to tie an animal's legs.

Prickly pear—a variety of cactus characterized by fleshy pads covered with both long sharp thorns and numerous clusters of fine needles. Wildlife and cattle eat some of these spiney but water-rich plants.

Prowling—the practice of mounted cowboys riding slowly through a pasture looking for signs of danger, illness, strays, or other problems in the pasture.

Quarter Horse—the preferred breed of horse for cattle work. These horses are characterized by stocky conformation and well-developed hindquarters and are known for quickness, agility, speed over short distances, and innate cow sense.

Quirt—a short whip used on horses, often of braided rawhide attached to a handle. These are rare among Texas cowboys, who use spurs instead.

Ranch horse—any horse raised, trained, and used on a ranch. A horse so called may well be of mediocre quality, that is, not highly trained or bred to rope, cut, or perform other specific work; often called a "using horse."

Ranch rodeo—contested rodeo events between teams of ranch cowboys as opposed to the traditional rodeo where professional cowboys compete as individuals.

Range cubes—a round or cubed piece of compressed protein-rich feed ranging from one-half inch to two inches or more in length and an inch or less in thickness. These cubes are usually dumped from mechanical feeders or sacks onto the ground.

Rattlesnake—a poisonous reptile common in Texas and the Southwest. The snake has a pair of fangs for injecting venom into its prey—usually mice and small rabbits—and "rattles" on its tail. The whirring of these rattles causes fear in anyone or any beast familiar with this deadly creature.

Rawhide—the untanned hide of an animal. The hair has usually been removed. Rawhide is very strong but softens when exposed to water.

Regional factors—those climatic, geographic, and cultural features peculiar to a part of the state. These include flora, fauna, soil types, and water sources typical of the area. All of these influence, even dictate, what kind of agricultural activities can be carried on there.

Rein—(n) the strip of leather or rope attached to the bit to guide the horse; (v) controlling a horse through directions given by the rider through pressure applied by the reins.

Remuda—the herd of horses, usually all geldings, kept on the ranch for riding.

Riding pasture—(v) checking cattle. See *prowling*.

Rigging—a general term to indicate the girth, tree, and other parts of the saddle used to hold the saddle in place on the horse's back.

River fever—an often fatal malady characterized by

a fever. It is common in low-lying areas of Southeast Texas.

Rodeo—(n) a competition between professional cowboys in such events as calf roping, bareback and saddle bronc riding, bull riding, and the like. See *ranch rodeo.*

Rolled cantle—that portion on the rear of the seat of the saddle. Only two or so inches high, it lays back or appears rolled over to the rear. Common in many cowboy saddles and on those designed for ropers so that the rider's leg can be easily passed over it.

Rope—a lariat or lasso of hemp, rawhide, or nylon.

Rope and drag—the method of roping a calf by the neck or heels and dragging it to a crew waiting to work the calf.

Rope corral—formed by a group of cowboys holding their lariats to form a loose pen for holding the remuda while a designated roper catches horses to be ridden for the day's work.

Rotating stock—moving livestock through a series of pastures, such as in cell grazing.

Roundup—a general term indicating the gathering of livestock for any purpose.

Rowel—the round, spikelike piece of metal attached to the shank of the spur. It is the part that touches the horse when the rider spurs the animal.

Running cattle—not necessarily meaning the moving of cattle at a run, the expression most often is equivalent to grazing cattle on the ranch, as in the expression "We are running eight hundred head of cattle."

Running iron—a straight or curved piece of steel attached to a steel handle. Rustlers used these to alter brands when stealing cattle. A heated cinch ring held by two green sticks can accomplish the same task and is less obvious when not in use.

Saddle—the throne of the cowboy. Many styles of saddles have been developed to suit specific needs.

Saddle horse—a horse trained to ride. A saddle horse may or may not be a trained cow horse.

Saddle house—a small building where saddles and related gear are kept. "Tack house" and "tack room" are equivalent but less frequently used terms.

Saddle tree—the wooden base to which the saddle leather is attached. The best are still made of wood (hence, tree) and covered with rawhide for strength.

Santa Gertrudis—a breed of cattle developed by King Ranch. Red in color, they show characteristics of Brahman and have proved particularly well adapted to the hot, humid areas along the coast.

Section—a unit of measurement equal to a square mile, or 640 acres of land.

Senepol—a breed of cattle, red in color, from the Virgin Islands.

Set—a number of calves, steers, cows, or other such groupings of approximately equal size, weight, or quality.

Shipping pasture—a pasture used only to hold the collected herd ready for shipping to market.

Shorthorn—a British breed popular in the early days of Texas ranching but in recent decades falling out of favor. Also called Durham.

Shotgun chaps—leather leg coverings with long, cylindrically shaped leg pieces resembling the barrel of a shotgun, hence the name. Long zippers are used to open and close each leg.

Simbrah—a crossbred bovine having Simmental and Brahman blood.

Simmental—a Swiss breed of cattle gaining in popularity in Texas. They crossbreed well with Herefords because of the red color.

Slicker—raincoat made for wearing while riding a horse.

Snaffle bit—a type of bit characterized by a large

ring on each side of the horse's mouth and a hinged bar across the mouth.

Soogan—a bedroll made up of a large tarpaulin intricately folded to provide protection from the cold.

Sorting cattle—the act of separating a herd into desired categories for any special purpose. For example, at shipping time, the marketable calves are separated from their mothers and may be sorted by weight to form the best set of calves for the buyer.

South Texas Diamond—the area described by Walter Prescott Webb with San Antonio on the north, Brownsville on the south, Indianola on the east, and Laredo on the west. In this part of South Texas, Webb said, ranching as we know it in the United States began.

Spur—a steel device attached to the cowboy's boot heel by a leather strap. The spur consists of a band of steel fitted around the heel of the boot, a shank attached to the rear of the band, and a rotating rowel attached to the end of the shank.

Spur leathers or straps—usually two leather straps to hold the spurs on the cowboy's boots. These are attached to the heel bands on each side and fastened by a buckle positioned on the outside of the foot.

Steer—a castrated male bovine.

Steer operation—the practice of stocking a ranch or part of a ranch with only steers, which are grazed for a time, often from two to six months, and then sold. See *cow-calf operation*.

Stovepipe boots—cowboy boots, the tops of which are straight without the traditional scallops in front and back.

Straight cantle—the raised rear part of the seat of the saddle. It rises some six inches or more rather than being laid over in the style called *rolled cantle*.

String—the horses assigned to a cowboy to ride in regular rotation. The string may include three to as many as twelve horses, depending upon the amount of riding done, the amount of feed given the horses, and the roughness of the terrain.

Swapping help—the practice of exchanging cowboys between ranches on heavy working days such as roundup or shipping days. See also *neighboring*.

Swell—the term to describe the front part of the saddle to which the horn is attached.

Tack—horse-related gear, including saddles, parts of the bridle, blankets, pads, stirrups, girths, latigos, and the like.

Tack room—room in a barn where saddles, bridles, and other riding gear is kept. Variant of *saddle house*.

Tapaderos—leather coverings over the front of a stirrup, sometimes in buckaroo culture with flaps extending twenty or more inches below the stirrup. Also called "taps" and "toe fenders."

Thoroughbred—a tall, slender, long-legged horse bred for racing. Sometimes crossbred with Quarter Horses to produce horses capable of speed over a long distance.

Tie-down—a leather strap attached to the bosal around the horse's nose running through a ring on the front of the breast collar, and attached to a ring on the front of the front girth. Used to keep a horse from tossing its head.

Toe fenders—See *tapaderos*.

Trap—a fenced area ranging in size from a few square yards to hundreds of acres in which livestock are kept. These are much smaller than the adjoining pastures and are designed to hold stock so that they can be observed closely or kept available for use.

Vaquero—the original mounted herder in North America. Originating in Mexico, his influence can be seen across cattle-raising areas of the West.

Wagon boss—on large ranches the man in charge

of the cattle-working crew. He gets his orders from the foreman or manager, who does not accompany the crew when doing the actual work. The term derives from the practice of managing the men who *work on the wagon* on a large ranch.

Windmill—a wind-powered device for pumping water. Mounted on towers, these are characterized by a large wheel with blades that catch the wind to turn the wheel.

Work cattle—(v) a general term to describe any of the several routine practices necessary to move through the annual cycle. This includes vaccinations, ear tagging, castration, branding, palpating, sorting, and the like.

Work on the wagon—an expression designating the routine of a group of men who live for varying periods around a chuck wagon. They sleep, eat, and spend their free time far from the headquarters. The chuck wagon is their home away from home.

Working table—a device constructed of metal sheeting and pipe to trap and hold an animal on its side for ease in working it. See also *calf cradle.*

Wrangle the horses—to round up and bring in the remuda to the corral or other location so that mounts can be selected for the work day. Sometimes called "jingle the horses."

Wrangler—the person whose job it is to bring in the remuda, separate the horses needed for the day's work, and feed the horses early enough so that they have finished eating before the men arrive before break of day to saddle up.

Wreck—a term for any of the numerous accidental falls, kicks, and other injuries common to working cowboys; a general term for anything going wrong.

Yearling—Calves or horses of either sex that are approximately one year old.

Bibliography

PRINTED SOURCES

Anders, Evan. "Canales, Jose Tomas." *The New Handbook of Texas*. Austin: Texas State Historical Association, 1996, 1:1953–1954.

Anderson, H. Allen. "Goodnight, Charles." *The New Handbook of Texas*. Austin: Texas State Historical Association, 1996, 3:240–243.

———. "Goodnight Ranch." *The New Handbook of Texas*. Austin: Texas State Historical Association, 1996, 3:245.

———. "Goodnight, Texas." *The New Handbook of Texas*. Austin: Texas State Historical Association, 1996, 3:244.

———. "JA Ranch." *The New Handbook of Texas*. Austin: Texas State Historical Association, 1996, 3:885–886.

———. "Whittenburg, James Andrew." *The New Handbook of Texas*. Austin: Texas State Historical Association, 1996, 6:950.

"Brown Ranch Ranked Ninth in Registrations." *Throckmorton Tribune*. Thursday, October 29, 1998, p. 1A.

Canales, Dawn. Letter to author, February 27, 2000.

Canales Family Archives. Premont, Texas.

Capps, Benjamin. *Sam Chance*. 1965. Reprint, Dallas: Southern Methodist University Press, 1987.

Carrizo Springs Javelina. Various undated clippings, Dimmit County Library vertical file.

Cellar, Carlos E. "Laredo." *The New Handbook of Texas*. Austin: Texas State Historical Association, 1996, 4:77–78.

Clayton, Lawrence. "A Cow Outfit Works Sheep." *Persimmon Hill* 25, no. 3 (autumn 1977): 75–78.

———. *Cowboys: Contemporary Ranch Life along the Clear Fork of the Brazos River*. Austin: Eakin, 1997.

———. "Frank Graham: A Cowboy at Heart." *Persimmon Hill* 23, no. 4 (winter 1996): 56–59.

———. *Historic Ranches of Texas*. Austin: University of Texas Press, 1993.

———. *Longhorn Legacy: Graves Peeler and the Texas Cattle Trade*. Abilene: Cowboy Press, 1994.

———. "Nig London, Throckmorton County Cowman." *West Texas Historical Association Year Book* 67 (1991): 94–100.

———. *Ranch Rodeos in West Texas*. Abilene: HSU Press, 1988.

Davies, Erin. "The Biggest Ranches." *Texas Monthly Magazine* 26 (August 1998): 118–125.

Dobie, J. Frank. *Tales of Old-Time Texas*. 1928. Reprint, Austin: University of Texas Press, 1984.

———. *A Vaquero of the Brush Country*. 1929. Reprint, Austin: University of Texas Press, 1981.
Douglas, C. L. *Cattle Kings of Texas*. 1939. Reprint, Austin: State House Books, 1989.
Emmett, Chris. *Shanghai Pierce: A Fair Likeness*. Norman: University of Oklahoma Press, 1953.
"Entire Community Mourns Capt. Jones' Death." *Beeville Bee-Picayune*. Reprint of article from 1905, n.p., n.d.
Erickson, John R. *Catchrope: The Long Arm of the Ranch Cowboy*. Denton: University of North Texas Press, 1994.
Farmer, Joan. "Budd Matthews, Texas—A Spur Track." Narrative to Support Application for Historical Marker, August 9, 1991.
———. "Remember When Column." *Albany News*. N.d., n.p.
Ford, John Salmon. *Rip Ford's Texas*. Ed. Stephen B. Oates. Austin: University of Texas Press, 1963.
Goddard, Dan R. "Photographs Honor Black Cowboys." San Antonio *Express-News*, January 24, 1988, p. 7H.
Graham, Joe S. *El Rancho in South Texas*. Denton: University of North Texas Press, 1994.
———. "Spanish and Mexican Origins of Ranching in Texas." *Journal Of Big Bend Studies* 10 (1998): 77–88.
Green Bob. "Nail Contributions 'Sizeable.'" *Albany News*, Fandangle Souvenir Section, June 1992, pp. 1–2B.
Greene, A. C. *A Personal Country*. New York: Knopf, 1969.
Grimes, Roy, ed. *300 Years in Victoria County*. Victoria, Tex.: *The Victoria Advocate*, 1968.
History of the Cattlemen of Texas. 1914. Reprint, with an introduction by Harwood Hinton, Austin: Texas State Historical Association, 1991.
Hollandsworth, Skip. "When We Were Kings." *Texas Monthly Magazine* 26 (August 1998): 112–117, 140–144.
Hunter, J. Marvin, ed. *The Trail Drivers of Texas*. 1924. Reprint, Austin: University of Texas Press, 1985.
"JA Ranch." http://www.ranches.org/JAranch.htm.
Jones, Kathryn. "Briscoe's Bounty." *Texas Monthly Magazine* 26 (August 1998): 126–127, 188–189.
———. "Shrinking Giant." *Texas Monthly Magazine* 26 (August 1998): 128–129, 147–149.
Jones, Nancy Baker. "Adair, Cornelia Wadsworth." *The New Handbook of Texas*. Austin: Texas State Historical Association, 1996, 1:22–23.
Kelton, Steve. *Renderbrook: A Century under the Spade Brand*. Fort Worth: Texas Christian University Press, 1989.
McCallum, Henry D., and Frances T. McCallum. *The Wire That Fenced the West*. Norman: University of Oklahoma Press, 1965.
McCoy, Dorothy Abbott. *Texas Ranchmen: Twenty Texans Who Help Build Today's Cattle Industry*. Austin: Eakin Press, 1987.
MacWhorter, William. "Arroyo Colorado." *The New Handbook of Texas*. Austin: Texas State Historical Association, 1996, 1:258.
Matthews, Sallie Reynolds. *Interwoven: A Pioneer Chronicle*. 1938. Reprint, College Station: Texas A&M University Press, 1982.
Nail, Reilly. *Per Stirpes: The John M. Nail Family in Texas, 1839–1995*. Abilene: Quality Printing, 1995.
Nelson, Barney. *The Last Campfire: The Life Story of Ted Gray, a West Texas Rancher*. College Station: Texas A&M University Press, 1984.
O'Connor, Louise S. *Cryin' for Daylight: A Ranching Culture in the Texas Coastal Bend*. Austin: Wexford Publishing, 1989.
Ormsby, Waterman L. *The Butterfield Overland Mail*. Ed. Lyle H. Wright and Josephine M. Bynum. San Marino, Calif.: Huntington Library, 1955.

"R. A. Brown." *Oklahoma Quarter Horse Breeders Association Magazine.* Undocumented clipping in Brown Ranch Archive.

Reeves, Frank. "The Reynolds Story." *Cattleman Magazine* 55, no. 7 (December 1968): 38–39, 54, 56, 58.

Reynolds, J. P. "Goodnight College." *The New Handbook of Texas.* Austin: Texas State Historical Association, 1996, 3:244.

Richardson, Rupert N. *Comanche Barrier to South Plains Settlement.* Ed. Kenneth W. Jacobs. 2d ed. Austin: Eakin, 1996.

Richardson, T. C. "Goodnight-Loving Trail." *The New Handbook of Texas.* Austin: Texas State Historical Association, 1996, 3:244–245.

Roach, Joyce Gibson. "Goodnight, Mary Ann Dyer." *The New Handbook of Texas.* Austin: Texas State Historical Association, 1996, 3:243–244.

"Straight from *Lonesome Dove.*" *North Dakota Horizons* 23, no. 2 (spring 1993): 25–27.

Taylor, Paul S. "Historical Note on Dimmitt County, Texas." *Southwest Historical Quarterly* 34, no. 1 (October 1930): 79–90.

Thompson, Jerry. "Cortina, Juan Nepomuccheno." *The New Handbook of Texas.* Austin: Texas State Historical Association, 1996, 2:343–344.

Tidwell, Laura Knowlton. *Dimmitt County Mesquite Roots.* [Austin]: Wind River Press, 1984.

Webb, J. R. "Chapters from the Frontier Life of Phin W. Reynolds." *West Texas Historical Association Year Book* 21 (1945): 110–143.

Webb, Walter Prescott. *The Great Plains.* Waltham, Mass.: Blaisdell, 1959.

Who's Who in Texas Today: A New Biographical Survey of Texas. Austin: Pemberton Press, 1968.

Woolley, Bryan. "A Legend Runs Through It." *Dallas Morning News,* June 29, 1998, pp. 1, 7–9F.

———. "Wide-Open Standoff." *Dallas Life Magazine,* January 16, 1994, pp. 8–12, 16, 18.

"Writer Describes Strong Character, Vivid Personality of W. W. Jones." Undocumented clipping, Jones family archive.

INTERVIEWS BY THE AUTHOR

Andrews, La Raine. By telephone, November 11, 1999.

Ball, Claudia. June 25–26, 1994.

Briscoe, Dolph, Jr. June 27, 1994.

Brown, Peggy. July 5, 1993.

Brown, Rob. July 5, 1993.

Brown, Valda. July 5, 1993.

Buckert, Kai. July 19, 1996.

Cameron, Darrell. By telephone, February 2, 2000.

Canales, Gus T. June 28, 1999.

Clayton, Donald H. August 19, 1995.

Driscoll, Mary Francis Johnson. March 13, 1992.

Graham, Frank. February 24, 1996.

Gray, Ted. August 18, 1996.

Green, Bob. January 10, 1991.

Hoskins, Curt. April 12, 1994.

Hoskins, Katy. April 12, 1994.

House, Elmer. August 19, 1995.

House, H. D. November 16, 1986.

Jones, Dick. June 23, 1994.

Jones, W. W., II. October 4, 1994.

Keefe, Joe. July 19, 1996.

Ledbetter, Morris. November 1, 1997.

Leech, Glenn. January 5, 1994.

London, Nig. November 20, 1990.

Moberley, Terry. July 10, 1988.

Moorhouse, Tom. October 14, 1999.

Northcutt, Bob. February 27, 1995.

O'Brien, Jay. By e-mail, February 7, 2000.

Palmer, Lee. November 12, 1999.

Pate, Jack. July 21, 1985.
Peacock, Benny. October 13, 1985.
Peacock, George. Numerous interviews between 1985 and 1995.
Pealer, Mary Fruzan Reynolds. October 13, 1993.
Ramírez, Chapo. August 18, 1996.
Reagan, Bob. August 19, 1995.
Reynolds, Jimbo. January 6, 1994.
Reynolds, Mary Joe. July 19, 1993.
Smith, Florene. January 22, 1993.
Waller, Robert. November 29, 1992.
Williams, Clayton, Jr. August 2, 1996.

OTHER INTERVIEWS

Mahoney, Dauris. Interviewed by Kurt House, 1972.